Lecture Notes in Physics

Edited by J. Ehlers, München, K. Hepp, Zürich, and
H. A. Weidenmüller, Heidelberg
Managing Editor: W. Beiglböck, Heidelberg

28

J. Ehlers, J. Ford, C. George, R. Miller,
E. Montroll, W. C. Schieve, J. S. Turner

Lectures in Statistical Physics

Advanced School for Statistical
Mechanics and Thermodynamics
Austin, Texas/USA

Edited by W. C. Schieve and J. S. Turner

Center for Statistical Mechanics and Thermodynamics
University of Texas, Austin, Texas/USA

Springer-Verlag Berlin Heidelberg GmbH 1974

ISBN 978-3-540-06711-5 ISBN 978-3-540-38006-1 (eBook)
DOI 10.1007/978-3-540-38006-1

Originally published by Springer-Verlag Berlin Heidelberg New York in 1974.

Library of Congress
Catalog Card Number 74-3671.

PREFACE

These articles have their origin in lectures presented at the third Advanced School for Statistical Mechanics and Thermodynamics held in Austin in 1971. The school was organized by the Center for Statistical Mechanics and Thermodynamics, Professor I. Prigogine, Director.

The lectures illustrate the remarkably wide scope of statistical physics, encompassing as they do the disciplines of physics, astrophysics, and biophysics. The article by Professor Schieve is an introduction to the formal theory of quantum non-equilibrium statistical mechanics from the point of view of the von Neumann equation. The recent "subdynamics" is discussed which shows that the general kinetic equation may in a sense be viewed as an exact dynamics. The quantum Boltzmann equation is simply obtained. Professor George continues the formal presentation of nonequilibrium statistical mechanics in a discussion of recent results within the general framework of quantum mechanical transformation theory. The "star unitary" transformation appropriate to large dissipative systems is examined and applied to the general kinetic equation to yield a simpler, more transparent form.*

Professor Ehlers then reviews kinetic theory in the context of Einstein's General Theory, discussing the justification of Boltzmann's ideas in curved space-time. The derivation of hydrodynamic equations and the adaptation of the Chapman-Enskog method are presented. Continuing the theme of astrophysics, Professor Miller discusses computer simulation of the formation and persistence of spiral structure in galaxies, concluding with comments on the difficulty of "thermodynamic" computer experiments with gravitating many-body systems.

Professor Montroll reviews one of the classic systems in statistical mechanics--the lattice. He discusses wave propagation in harmonic,

---

*This article is a reproduction of a reprint available at the school.

anharmonic, and defect lattices. Important questions of ergodicity are raised in an analysis of permanent waves (solitons) in the one dimensional chain. These questions are explored further in Professor Ford's article on computer solutions of few-particle nonlinear oscillator systems. Such simulations illustrate visually the violation of the Kolmogorov, Arnold, Moser criteria for orbital stability and the subsequent transition to stochastic behavior and exponentiating orbits (C-systems). The incredibly complex evolution of a simple classical two-oscillator system near hyperbolic fixed points is presented in the mathematical context of area-preserving mappings.

Turning in the final article to recent developments in nonequilibrium thermodynamics, Dr. Turner discusses instabilities which may occur in nonlinear systems maintained far from thermal equilibrium and possible evolution beyond instability to new types of states characterized by organization in space, time, or function. By way of illustration a few model chemical systems are presented, and the general principles applied to selected aspects of the problem of biological order.

Finally we thank the National Science Foundation, the University of Texas at Austin, General Motors Corporation, and the Ford Motor Company, whose joint sponsorship of the School made this volume possible.

William C. Schieve

Jack S. Turner

CONTENTS

ASPECTS OF NON-EQUILIBRIUM QUANTUM STATISTICAL MECHANICS: AN INTRODUCTION

William C. Schieve
University of Texas at Austin

CHAPTER I. INTRODUCTION

A. Remarks

These lectures are an introduction to non-equilibrium statistical mechanics. The emphasis will be on quantum statistical mechanics since the microscopic world is quantum mechanical and classical statistical mechanics is but an interesting limit. The fundamental tool of quantum statistical mechanics is the density operator, ρ, whose time evolution is governed by the von Neumann equation. Its matrix elements, $\langle\alpha|\rho|\beta\rangle$, describe probabilities as well as correlations in the system. It was first introduced by von Neumann [1] and Landau [2] in the early years of quantum mechanics.* In a sense a more general quantum mechanics may be formulated by means of this density operator allowing for a description of "mixtures" as well as "pure states". It also has the added appeal of being more directly observable in that it contributes linearly to expectation values rather than in a "quadratic" fashion as the wave function.

Mixtures are states of incomplete quantum mechanical knowledge. They form the foundation of quantum statistical mechanics. Just as in classical statistical mechanics in most experiments in a many-body system it is a practical impossibility to determine the maximal quantum mechanical knowledge (the pure state) initially [5].** It must

---

*See also references [3] and [4].

**It should be noted that Tolman discusses this point in the context of equilibrium statistical mechanics and thus the initial value character is obscured.

be emphasized that this is true even in the quantum mechanics of systems with few degrees of freedom. As an example in scattering of particles with intrinsic spin it is not always possible to determine the initial spin orientation and thus a mixture state is the appropriate incoming state [6]. Another example is a state prepared by two different Stern-Gerlach apparatuses using different sources. The recombined beam is in a mixture state in the internal variables. The weighting parameters are the fractional intensities of the two beams. In contrast a pure state may be prepared by one Stern-Gerlach apparatus and source.

The very terminology "mixture" is perhaps misleading as it is normally thought of as an incoherent superposition of pure states with arbitrary weighting. This superposition is not unique. It is better perhaps, as emphasized by Fano [4], to adopt the density operator itself as the definition of the "state", and to incorporate the information (maximal or not) at the initial time into an initial condition for the von Neumann equation. This uncertainty in the initial condition leads naturally to the notion of ensembles -- identical quantum mechanical systems prepared in the same way. Without the maximal quantum mechanical knowledge there may be arbitrary parameters to assign initially. For instance, in the spin 1/2 scattering experiment alluded to above $\rho = \frac{1}{2}(1 + \underset{\sim}{\sigma} \cdot \underset{\sim}{P})$. Here $\underset{\sim}{\sigma}$ is the Pauli spin matrices, and $\underset{\sim}{P}$ the polarization. For $\underset{\sim}{P} = 0$ we have equal mixture of $|1/2\rangle$ and $|-1/2\rangle$ states. A pure state ($\rho^2 = \rho$) results if $|\underset{\sim}{P}| = 1$.

The principle aim of non-equilibrium statistical mechanics is to explain in a fundamental way the irreversible approach of many-body systems to equilibrium, and to evolve practical kinetic equations to describe such phenomena as relaxation or transport properties in matter in its various phases: solid, liquid and gaseous. In discussing irreversibility the object is to show under what conditions time asymptotic kinetic equations may be obtained which describe the approach of an isolated system to thermodynamic equilibrium. A central theme in this is the search for an H theorem analogous to that of Boltzmann [7], $\dot{H} \leq 0$ which is the microscopic statement of the second law of thermodynamics ($H = -kS$, S being the entropy and k Boltzmann's constant). It is easily seen that for a discrete spectrum the von Neumann equation gives a $\langle\alpha|\rho|\beta\rangle$ which is almost-periodic, thus exhibiting the analogue of classical Poincaré recurrences. This of course leads to a quantum form of Zermelo's paradox [8]. It is well known that such recurrence times become super cosmological (10^{10^N} sec.) as the number of degrees of freedom, N, becomes large. None the less they are a difficulty for the theory. A way of mathematically removing them has been to obtain a continuous spectrum via the thermodynamic limit, $N\to\infty$, $V\to\infty$, N/V = constant. This, unfortunately, leads to new problems similar to those which exist in quantum field theory [9]. While Fock space is an adequate tool for treating non-interacting fields (particles), simple models [10] show that with the presence of the interaction such maladies appear as infinite renormalization, impossibility of defining a proper total Hamiltonian, and the non-existence of a ground state. The time revolution is also not unitarily implemented. For the more rigorously minded a possible choice is to adopt the interesting algebraic methods [10] whose purpose is to supersede the Fock space difficulties. However, for the purpose of discussing quantum kinetics this is not yet a useful tool, and we shall use here more familiar techniques which have not yet been cast into a rigorous mathematical form but which have led to useful new results. Useful examples of kinetic equations are the Peierls equation [11] governing phonon transport in solids, the Waldman-snider equation for a dilute gas with internal degrees of freedom, and the optical pumping equation for atoms interacting with radiation [12]. Systems with man degrees of freedom such as these ($N \approx 10^{23}$) have the important property of being describable on the macroscopic level by simple quantities such as particle number, velocity, entropy, entropy production and fluxes of heat, electric current, etc. Because of the nature of these observables not all the elements of $\langle\alpha|\rho|\beta\rangle$ are needed. Generally a partial trace or a projection of ρ gives sufficient information. For a gas it is sufficient to know reduced 1-particle density operators. We shall see that this fundamental simplifying property is reflected in the kinetic equations.

The main aim of this introduction is to provide the reader with the theoretical foundations used in obtaining quantum kinetic equations from the von Neumann equation. In part B of this section we will introduce the von Neumann equation and the postulates of quantum statistical mechanics. In section II the generalized master equation (G.M.E.) [13,14] approach will be discussed briefly. To illustrate the results we will discuss the solution to the simple Friedrichs model [15] illustrating the asymptotic properties. In the context of section II the article by R. Balescu in the first of this series is relevant [16].

In section III we will discuss the newer and more elegant alternative to the G.M.E., the concept of "subdynamics" first introduced by Prigogine, George and Henin [17,18,19]. In this alternative formulation an idempotent operator, $\tilde{\Pi}$, may be introduced which separates out the kinetic part of the density operator. This part, $\tilde{\Pi}\rho$, is governed by a generalized kinetic equation at <u>all times</u>, which is independent of its complement, $(1-\tilde{\Pi})\rho$. Thus, the general kinetic equation appears in a new light and in a more central way. This section is related to the lectures of C. George in the subsequent article in this volume.* There the motive is to go further introducing the star unitary transformation

---

*Hereafter, referred to as GI.

and obtain a "physical particle representation" - a transformed generalized kinetic equation.

In section IV we derive the quantum Boltzmann equation [20] and explore its relationship to scattering theory [6]. For simplicity we will use the B.B.C.Y.K. hierarchy (Bogoliubov, Borne, Green, Kirkwood, Yvon) for reduced density operators [21] to obtain it. As discussed there is has not yet been possible to construct an appropriate realization of a projection operator which will in a concise manner yield the Boltzmann equation for spatially inhomogeneous systems. The hierarchy will be treated in a manner analogous to the G.M.E. .

B. Fundamentals

In statistical mechanics the state of the system is described by density operator (density matrix), ρ, having the properties

$$\mathrm{Tr}\rho = 1; \ \langle\alpha|\rho|\alpha\rangle > 0, \ \rho^{\dagger} = \rho \ . \tag{1.1}$$

After Emch [22] (for finite degrees of freedon), we may introduce a lioville space L, a linear space, of which the density operator is an element. L is spanned by the quantum mechanical observables A, B, ... The inner product is defined by

$$(A,B) = \mathrm{Tr}A^{\dagger}B. \tag{1.2}$$

The expectation value (ensemble average) of A is taken to be

$$\langle A\rangle_t = (\rho(t),A) = \mathrm{Tr}\rho(t)A. \tag{1.3}$$

For finite number of degrees of freedom A might be taken to have a Schmidt norm [22] and L to be a Hilbert space. However, in physical applications unbounded operators are used and the important thing is (2.1) exists for "well behaved" A. This assumption is in fact intrinsic to the operator manipulations of sections II and III.

We may introduce superoperators [23], $\mathcal{A}$, $\mathcal{B}$, $\mathcal{C}$, which map the Liouville space onto itself. They are operators performing linear transformation of quantum operators (vectors) into themselves. We may define the adjoint superoperator, $\mathcal{A}^\dagger$ by

$$(A, \mathcal{A}^\dagger B) = (\mathcal{A}A, B) \tag{1.4}$$

and with this hermitian ($\mathcal{A}^\dagger = \mathcal{A}$) and unitary ($\mathcal{A}\mathcal{A}^\dagger = \mathcal{A}^\dagger\mathcal{A} = 1$) super operators. The most important unitary superoperator is exp(-iLt) where the quantum Liouville operator

$$L = [H,] \quad (L^\dagger = L) \tag{1.5}$$

is the generator of the mapping of $\rho(0)$ into $\rho(t)$,

$$\hbar = 1 \qquad \rho(t) = \exp(-iLt)\rho(0), \tag{1.6}$$

being the solution to the von Neumann equation [1,2]

$$i\dot{\rho}(t) = L\rho(t). \tag{1.7}$$

Relationships (1.1), (1.3) and (1.7) are the fundamental assumptions of quantum non-equilibrium statistical mechanics. The von Neumann equation is equivalent to the Schrödinger equation or Heisenberg equations of motion of the more familiar quantum mechanics.

For calculation it is necessary to adopt rules for calculating matrix elements in the ordinary quantum Hilbert space of states, $|\alpha\rangle$. For an $|\alpha\rangle$ which are eigenstates of H° for instance we may form a dyad operator $|\alpha\rangle\langle\alpha^1|$ and generalize Dirac's notation* forming a vector in the super space

$$\begin{aligned} |\alpha\alpha^1\rangle &\equiv |\alpha\rangle\langle\alpha^1| \\ \langle\alpha\alpha^1|\beta\beta^1\rangle &\equiv \langle\alpha|\beta\rangle\langle\alpha^1|\beta^1\rangle . \end{aligned} \tag{1.8}$$

Then, $|\alpha\alpha^1\rangle$ are eigenstates of L°

---

*This form of the tetradic representation of super operators has proven useful in the discussion of decay systems, J. W. Middleton (unpublished work).

$$L^\circ|\alpha\alpha^1\rangle = [H^\circ, |\alpha\rangle\langle\alpha^1|] = (\alpha - \alpha^1)|\alpha\alpha^1\rangle \tag{1.9}$$

with eigenvalue $(\alpha - \alpha^1)$. We may represent a quantum observable

$$A = \sum_{\alpha\alpha^1} \langle\alpha\alpha^1|A\rangle|\alpha\alpha^1\rangle \tag{1.10}$$

where

$$\langle\alpha\alpha^1|A\rangle \equiv \langle\alpha|A|\alpha^1\rangle \ . \tag{1.11}$$

A super operator, $\mathcal{A}$, may be represented by a superposition of dyads in the super space

$$\mathcal{A} = \sum_{\alpha\alpha^1\beta\beta^1} \langle\alpha\alpha^1|\mathcal{A}|\beta\beta^1\rangle|\alpha\alpha^1\rangle\langle\beta\beta^1| \tag{1.12}$$

where the

$$\mathcal{A}_{\alpha\alpha^1,\beta\beta^1} \equiv \langle\alpha\alpha^1|\mathcal{A}|\beta\beta^1\rangle$$

is a tetrad, the tetradic representation of $\mathcal{A}$ [4]. The rules for multiplying tetrads with tetrads and tetrads with usual matrices can be easily obtained. They are

$$\mathcal{A}_{\alpha\alpha^1,\beta\beta^1} = \sum_{\delta\delta^1} \mathcal{B}_{\alpha\alpha^1,\delta\delta^1}\mathcal{C}_{\delta\delta^1,\beta\beta^1}$$

and (1.13)

$$A_{\alpha\alpha^1} = \sum_{\delta\delta^1}{}^1 \mathcal{B}_{\alpha\alpha^1,\delta\delta^1}C_{\delta\delta^1} \ .$$

Of particular importance is the tetradic representation of the commutator operator L.

$$L = [H,\ ] = \sum_{\substack{\alpha\alpha^1\\ \beta\beta^1}} \langle\alpha\alpha^1|[H,\ ]|\beta\beta^1\rangle|\alpha\alpha^1\rangle\langle\beta\beta^1| \ ,$$

where we define

$$L_{\alpha\alpha^1,\beta\beta^1} \equiv \langle\alpha|[H, |\beta\rangle\langle\beta^1|]|\alpha^1\rangle = \langle\alpha|H|\beta\rangle\langle\beta^1|\alpha^1\rangle - \langle\beta^1|H|\alpha^1\rangle\langle\alpha|\beta\rangle \ . \tag{1.14}$$

Also we will frequently need the matrix representation of ρ,

$$\rho = \sum_{\alpha\beta} \langle\alpha\beta|\rho\rangle|\alpha\beta\rangle \ , \tag{1.15}$$

where $\langle\alpha\beta|\rho\rangle = \langle\alpha|\rho|\beta\rangle$. The diagonal elements have the meaning of the probability that the system is in eigenstate $|\alpha\rangle$. A simple example illustrating this is the free particle in a 1-dimensional box of length L with periodic boundary conditions. Assuming the canonical distribution, $\rho = \exp(-\beta H)$, the matrix elements in the position representation are

$$\langle X|\rho|X'\rangle = \left(\frac{m}{2\pi\beta}\right)^{\frac{1}{2}} \exp\left(-\frac{m}{2\beta}\,(X - X')^2\right).$$

The probability that the particle is at X is $\left(\frac{m}{2\pi\beta}\right)^{\frac{1}{2}}$. One sees the average value of the position and momentum is zero and the average kinetic energy is $\frac{1}{2}\beta = \frac{kT}{2}$. The off-diagonal elements have no simple physical meaning.

Finally, we remark that in subsequent sections we will drop the script notation for superoperators. Otherwise, a whole notational change from the literature relevant to sections II and III would be necessary.

CHAPTER II. THE GENERALIZED MASTER EQUATION

Generally experiments upon many-body systems give us reduced information--less detailed than the full density operator. For instance we may know the probability that the system exists in some state $|\alpha\rangle$. One means of separating out this reduced information is to introduce the projection operator, P, first introduced by Nakajima [24] and independently by Zwanzig [14], and used by Prigogine, George Henin and Rosenfeld [18] following the work of M. Baus.*

Define with the super-dyad $|\alpha\alpha\rangle\langle\alpha\alpha|$

$$P = \sum_{\alpha} |\alpha\alpha\rangle\langle\alpha\alpha| \ . \tag{2.1}$$

It is easily seen from Equation (1.15) that

$$P\rho = \sum_{\alpha} \langle\alpha\alpha|\rho\rangle |\alpha\alpha\rangle \ , \tag{2.2}$$

and hence the non-vanishing matrix elements

$$\langle\alpha|P\rho|\alpha^1\rangle = \langle\alpha\alpha|\rho\rangle\langle\alpha|\alpha^1\rangle$$

give the probability that system is in state $|\alpha\rangle$, the eigenstate of the Hamiltonian (unperturbed!) $\overset{\circ}{H}$. It is easily seen that

$$\begin{aligned} P^2 = P \quad , \quad P^{\dagger} = P \\ PL^{o} = L^{o}P = 0 \ . \end{aligned} \tag{2.3}$$

With this we separate the density operator

$$\rho = P\rho + (1 - P)\rho \equiv \rho_o + \rho_c \ . \tag{2.4}$$

Here, following Balescu [16] we adopt the physical point of view suggested by the perturbation theoretic study of the dynamics of correlations [13] and call ρ_o the uncorrelated (correlation vacuum) part of the density operator and ρ_c the correlations.

---

*M. Baus concisely formulated classically the work of the Brussels group in terms of projection operators. See Acad. Roy. Belg. Bull. Cl. Sci. 1967, 53, 1291.

We immediately write the von Neumann equation as

$$i\dot{\rho}_o = L_{oo}\rho_o + L_{oc}\rho_c \tag{2.5a}$$

$$i\dot{\rho}_c = L_{cc}\rho_c + L_{co}\rho_o \tag{2.5b}$$

where

$$L_{oo} = PLP$$

$$L_{co} = (1 - P)LP$$

etc.

The formal solution to Equation (2.5b) is

$$\rho_c(t) = -i\int_0^t d\tau \, \exp(-iL_{cc}\tau)L_{co}\rho_o(t-\tau) + \exp(-iL_{cc}t)\rho_c(o). \tag{2.6}$$

We should remark that implicit to this solution is a causality condition mentioned in the introduction which chooses the causal Green's function [25,26] of Equation (2.5b). The discussion is somewhat subtle [26]* since it is only in the thermodynamic limit and with the assumption of a continuous spectrum that the solution of the homogeneous von Neumann-like equation

$$i\dot{\rho}_c = L_{cc}\rho_c$$

and the time reversed (adjoint) equation become the same. Using Equation (2.6) in Equation (2.5a) we obtain the generalized master equation (G.M.E.)

$$\dot{\rho}_o + iL_{oo}\rho_o = \int_0^t G(\tau)\rho_o(t-\tau)d\tau - \mathcal{D}(t)\rho_c(o) \tag{2.7}$$

---

*The discussions in references [25] and [26] are for the classical case but the considerations here are the same where

$$G(\tau) = -L_{oc}\exp(-iL_{cc}\tau)L_{co} \tag{2.8a}$$

and

$$\mathcal{D}(t)\rho_c(o) = -iL_{oc}\exp(-iL_{cc}t)\rho_c(o) \ . \tag{2.9a}$$

This was first obtained using perturbation theoretic arguments by Prigogine and Resibois [27], and with these projection methods by Nakajima [23] and Zwanzig [14].

The first term on the right hand side is the characteristic non-Markoffian (memory of collision) term, and the last term describes the effect of the initial correlations upon the vacuum of correlations. The whole point is to judiciously choose P so that one can prove or have some physical expectation that this term will not have an effect after a collision duration time, which is much shorter than the relaxation time. Very few calculations have been made to actually prove this is so, since this leads to the difficult question of the characterization of $\rho_c(o)$. Jones [28] showed the disappearance of this for the case of classical particles interacting weakly with random scattering centers and assuming finite range initial correlation. Sewell [29] has used a course-grained projector characterizing macroscopic measurement. Assuming as a result of macroscopic measurement an initial state such that the vectors within the course-grain cell are equally probable and have random phases he then shows that the effect of the initial correlations vanishes identically. With this term missing Equation (2.7) becomes a closed equation for the relevant part of the density operator, $\rho_o(t)$. In the next section we will show how the "subdynamics" [17,19] can achieve a closed equation in a much more direct fashion.

We might mention that for the analysis of the solution of

Equation (2.7) it is most convenient to work with Laplace transforms [25].* Then, the collision superoperator, G(t), and "destruction" superoperator, $\mathcal{D}(t)$, are

$$G(t) = (2\pi i)^{-1}\int_c dz\ \exp(-izt)\psi(z) \tag{2.8b}$$

$$\mathcal{D}(t)\rho_c(o) = (2\pi)^{-1}\int_c dz\ \exp(-izt)\mathcal{D}(z)\rho_c(o) \tag{2.9b}$$

where

$$\psi(z) = L_{oc}\mathcal{P}(z)L_{oc} \tag{2.10}$$

$$\mathcal{D}(z) = L_{oc}\mathcal{P}(z) \tag{2.11}$$

$$\mathcal{P}(z) = (1 - P)(z - L_{cc})^{-1}(1 - P) \quad . \tag{2.12}$$

In terms of these the formal solution to the G.M.E. is

$$\begin{aligned}\rho_o(t) &= -(2\pi i)^{-1}\int_c dz\ \exp(-izt) \\ &\cdot (z - L_{oo} - \psi(z))^{-1}[\rho_o(o) + \mathcal{D}(z)\rho_c(o)] \ .\end{aligned} \tag{2.13}$$

Let us consider the asymptotic limit and the idea of time scaling already alluded to which was first introduced by Bogoliubov [30]. We make the fundamental assumption of the existence of two separated time scales, a collision time, τ_c, and a relaxation time, $\tau_R > \tau_c$. The one, τ_c, describes the decay of G(t) and $\mathcal{D}(t)\rho_c(o)$. It may be thought of as related to a duration of a collision. The other, τ_R, describes the relaxation of the system to equilibrium. We will not in this section be concerned with hydrodynamic effects so that an even longer characteristic hydrodynamic time, τ_h, will not enter. We take the time

---

*The contour c then runs from $-\infty + \delta$ to $\infty + \delta$, $\delta > 0$ above all singularities.

scales to be separated and assume

$$\lim_{t\to\tau_R} \left\{ \begin{matrix} G(t)A \\ \mathcal{D}(t)\rho_c(o) \end{matrix} \right\} = 0 \quad . \tag{2.14}$$

Here A is a suitable density operator whose matrix elements with the "irreducible" time evolution operator, $\exp(-iL_{cc}t)$, in the thermodynamic limit obey the Lipschitz condition [13, 25]. In Laplace transforms this takes the form of the assumption that the matrix elements of the irreducible operators $\psi(z)A$ and $\mathcal{D}(z)\rho_c(o)$, are regular functions of z in the neighborhood of $z = 0$. This is the statement of the strong asymptotic hypothesis. Certainly it may be weakened since one always has in mind that Equation (2.14) will be taken as an ensemble average with suitable chosen (time smoothed - macroscopic) observables. Then, the asymptotic hypothesis can be viewed as approximately true in the sense of weak convergence [31].

For times of the order of $t \approx \tau_R$ we may neglect the effect of initial correlations dropping the second term on the right hand side of Equations (2.6) and (2.7). Denoting the asymptotic density operator by $\tilde{\rho}$ we obtain

$$\dot{\tilde{\rho}}_o + iL_{oo}\tilde{\rho}_o = \int_o^\infty G(\tau)\tilde{\rho}_o(t - \tau)d\tau \tag{2.15}$$

$$\dot{\rho}_c = -i\int_o^\infty \exp(-iL_{cc}\tau)L_{co}\tilde{\rho}_o(t - \tau)d\tau \quad . \tag{2.16}$$

Let us further assume that Equation (2.11) is governed by an asymptotic time displacement superoperator, Θ, such that

$$\tilde{\rho}_o(t) = \exp(-i\Theta t)\tilde{\rho}_o(o) \quad . \tag{2.17}$$

This transformation is motivated by the search for exponentially decaying solutions (even approximate). From Equation (2.15) we obtain a non-linear operator integral equation

$$\Theta = L_{oo} - i\int_0^\infty d\tau G(\tau)\exp(i\Theta\tau)^* \quad . \tag{2.18}$$

Further, we have

$$\tilde{\rho}_c(t) = C\tilde{\rho}_o(t) \ , \tag{2.19}$$

where

$$C = -i\int_0^\infty d\tau \ \exp(-iL_{cc}\tau)L_{co} \ \exp(i\Theta\tau) \quad . \tag{2.20}$$

From Equation (2.18) and (2.20) we have

$$\Theta = L_{oo} + L_{oc} \ C \ . \tag{2.21}$$

The asymptotic correlations are related to the correlation vacuum by the operator C. This is a generalization of the Bogoliubov functional ansatz.

The operator integral equation for Θ may be formally solved by iteration based upon setting $\Theta = 0$ on the right hand side of Equation (2.18). For the case $L_{oo} = 0$ in the lowest order

$$\Theta^{(0)} \equiv \psi(+io) = i\int_0^\infty G(\tau)d\tau \ .$$

The operator Θ, and the higher iteration scheme was first introduced by Resibois [32]. We will not write it down in general but mention it in context of a simple example shortly. In the next section we will see that the asymptotic and seemingly approximate Equations (2.17) and (2.20) are in fact exact and may be viewed as fundamental to the theory. Before turning to this point let us illustrate some features of these formal operators by a simple model.

---

*In the earlier literature such as references [13], [17] and also in GI Θ is denoted by $\Omega\psi$.

Let us consider the model of a decay system of Friedrichs [15,33]. Choose a complete set of states consisting of a single state $|E\rangle$ and a continuum $|\omega\rangle$. We require

$$\langle E|E\rangle = 1 \quad \langle\omega|\omega^1\rangle = \delta(\omega-\omega^1) \quad \langle E|\omega\rangle = 0 \ , \tag{2.22}$$

the unperturbed Hamiltonian being

$$H^\circ = E|E\rangle\langle E| + \int d\omega\omega|\omega\rangle\langle\omega| \ . \tag{2.23}$$

The interaction takes place only between the single discrete state and the continuum, such that

$$\langle\omega|H'|E\rangle = \langle E|H'|\omega\rangle^* = V(\omega) \ . \tag{2.24}$$

In this case there are only few non-zero matrix elements of L', Equation (1.14), in the unperturbed basis. We have

$$\begin{aligned} L'_{\omega\mu E\nu} &= L'^*_{E\nu\omega\mu} = V(\omega)\delta(\mu-\nu) \\ L'_{\nu E\mu\omega} &= L'^*_{\mu\omega\nu E} = -V(\omega)\delta(\mu-\nu) \\ L'_{\omega EEE} &= L'^*_{EE\omega E} = V(\omega) \\ L'_{EEE\omega} &= L'^*_{E\omega EE} = -V(\omega) \ . \end{aligned} \tag{2.25}$$

Let us choose as the projector

$$P = |EE\rangle\langle EE| \tag{2.26}$$

projecting out of the density operator those diagonal matrix elements giving the probability that the discrete state is occupied.

Using the definition of the superoperator and the resolvant identity

$$(z-L)^{-1} = (z-L^0)^{-1} + (z-L^0)^{-1}L'(z-L)^{-1} \tag{2.27}$$

we have

$$P(z) = P^0(z)[1 + L'_{cc}P(z)] \tag{2.28}$$

where

$$P^0(z) = (1-P)(z-L^0_{cc})^{-1}(1-P) \quad . \tag{2.29}$$

Matrix elements of Equation (2.28) give a coupled set of integral equations for the elements of $P(z)$. These equations are soluble having factorable kernels. We will not go into details. The reader should consult the paper of Middleton and Schieve [15]. From $P(z)$ all the important operators, $\psi(z)$ and $\mathcal{D}(z)$, Equations (2.10) and (2.11) may be obtained.

The model is further simplified by assuming an explicit form for the interaction and taking the lower bound on the continuous spectrum as $-\infty$ [34]. We take

$$|V(\omega)|^2 = \frac{g^2\gamma^3}{4\pi}[(\omega - E)^2 + \gamma^2]^{-1} \tag{2.30}$$

where $g^2 = 4\pi \frac{\lambda}{\gamma}$ is a dimensionless height, λ, to width, γ, ratio. We obtain

$$\psi(z)_{EEEE} = \frac{2\alpha\beta(z + 2i\gamma)}{(z + i\mu_+)(z + i\mu_-)} \tag{2.31}$$

$$\mathcal{D}(z)_{EEE\omega} = \frac{-V(\omega)[z + \omega - E + i\gamma + \tfrac{1}{2}\psi(z)_{EEEE}]}{(z + \omega - E + i\alpha)(z + \omega - E + i\beta)}$$

$$\mathcal{D}(z)_{EE\omega E} = \frac{V^*(\omega)[z + E - \omega + i\gamma + \tfrac{1}{2}\psi(z)_{EEEE})]}{(z + E - \omega + i\alpha)(z + E - \omega + i\beta)} , \tag{2.32}$$

$$\mathcal{D}(z)_{EE\mu\nu} = (z + \nu - \mu)^{-1}[V^*(\mu)\mathcal{D}(z)_{EEE\nu} - V(\nu)\mathcal{D}(z)_{EE\mu E}]$$

where the characteristic constants are

$$2\alpha = \gamma[1 - (1 - g^2)^{\frac{1}{2}}]$$

$$2\beta = \gamma[1 + (1 - g^2)^{\frac{1}{2}}] \qquad (2.33)$$

$$2\mu_{\pm} = 3\gamma \pm \gamma(1 - 2g^2)^{\frac{1}{2}} \ .$$

Matrix elements of the collision operator, $\psi(z)$, are analytic in the upper half-plane and in the neighborhood of $z = 0$. Singularities characterizing the decay times of the system lie in the lower half plane. The collision kernel decay is characterized by $\mu_{\pm}^{-1}$. For very weak coupling, $g^2 << 1$, this is equal to γ^{-1}, the breadth of the interaction. Similar results hold for $\mathcal{D}(z)\rho_c(o)$ for suitable $\rho_c(o)$. We will discuss this more fully later. In the general analysis of many-body systems, Prigogine, et al. [35] have described a "dissipative" system by the necessary condition

$$\lim_{\varepsilon \to o} \psi(+ i\varepsilon) \neq 0 \ . \qquad (2.34)$$

In this model $\psi(+ 0) = 0$ if (a) $\lambda = 0$ or $\gamma = 0$ so that the interaction vanishes almost everywhere or (b) $\lambda \to \infty$, $\gamma \to 0$ such that $\lambda\gamma$ = const, then the interaction is proportional to $\delta(\omega - E)$. (a) is the physical situation of no interaction between the unperturbed states analogous to no collisions in an atomic gas. In (b) the interaction is singular only one state of the continuum, $|\omega\rangle$, may cause transitions. The continuous spectrum is not fully utilized.

Let us consider the solution, Equation (2.9). We will assume the random phase initial conditions

$$\rho_{EE}(0) = 1$$

$$\rho_c(0) = 0 \ . \qquad (2.35)$$

We have

$$\rho(t)_{EE} = -(2\pi i)^{-1} \int_C dz \, \frac{\exp(-izt)(z + i\mu_+)(z + i\mu_-)}{(z + 2i\alpha)(z + 2i\beta)(z + i\gamma)} \tag{2.36}$$

and, thus, by the residue theorem

$$\rho(t)_{EE} = (\beta-\alpha)^{-2}[\beta^2 \exp(-2\alpha t) + \alpha^2 \exp(-2\beta t) - 2\alpha\beta \exp(-\gamma t)]. \tag{2.37}$$

For weak coupling, $g^2 < 1$, α and β are real. Thus, for $t \gg \gamma^{-1}$

$$\rho(t)_{EE} = \beta^2(\beta-\alpha)^{-2} \exp(-2\alpha t) \quad . \tag{2.38}$$

Whereas, for $g^2 > 1$ α, β have imaginary parts and the solution is

$$\rho(t)_{EE} = \frac{\exp(-\gamma t)}{G^2} \, [\gamma G \sin Gt + \tfrac{1}{2}(G^2-\gamma^2)\cos Gt + \tfrac{1}{2}(G^2+\gamma^2)] \tag{2.39}$$

where

$$G = \gamma(g^2-1)^{\frac{1}{2}} \quad .$$

The most striking point is the transition from simple exponential decay in weak coupling to damped oscillatory decay in the strong coupling regime.

In very weak coupling the ratio of the collision time, γ, to the relaxation time $\tau_R = (2\alpha)^{-1}$, is

$$\frac{\tau_c}{\tau_R} = [1 - (1-g^2)^{\frac{1}{2}}] \approx \frac{g^2}{2} \quad ,$$

being less than unity and directly proportional to the coupling constant g^2. However, in strong coupling this is no longer the case, the decay proceeds with damped oscillations and strictly speaking there is no relaxation time. If we identify τ_R with the envelope of the decay, γ^{-1}, then $\frac{\tau_c}{\tau_R} \approx 1$. Analytical models showing similar features have been discussed by Van Hove-Verboven [36] and Haubold [37]. The Friedrichs model has many of the features of the true many-body system. Of course there is no equilibrium; the probability of the discrete states goes to zero as $t \to \infty$. Also, the assumption of the continuous spectrum $|\omega>$ means the system is already in the thermodynamic limit. The analogue to the usual limit in this model would be to normalize to finite volume obtaining discrete $|\omega_n>$ and then take $V = >\infty$ to obtain the continuous spectrum.

Grecos and Prigogine [38] have discussed other initial conditions for this model. It is easy to see by writing out the matrix elements that contributions to the solution of the G.M.E. from $\mathcal{D}(z)\rho_c(0)$ are of the form

$$\int d\omega \, \frac{V(\omega)g(\omega)}{\omega - z} \quad .$$

This is a Cauchy integral and providing $g(\omega)$, the matrix elements of $\rho_c(0)$ obey the Lipschitz condition [25]. It may be shown that this contribution vanishes as $\frac{1}{t}$ as $t \to \infty$. This is really not strong enough. If we take the spectrum over the whole real axis, as above, and if $V(\omega)$ and $g(\omega)$ may be analytically continued, then the contribution vanishes exponentially. However, this contribution will not vanish if $g(\omega)$ is a distribution.

Let us discuss the iteration of Equation (2.18) mentioned on page 14 . For this simple model it becomes [32]

$$\Theta = \sum_{n=1}^{\infty} \frac{1}{n!} \left[\frac{d^{n-1}}{dz^{n-1}} \psi^n(z)\right]_{z=0} \quad , \qquad (2.40)$$

and one can investigate the convergence of the series numerically with the aid of the Lagrange theorem [15]. It is found that for $g^2<1$ convergence can be proved. The iteration of Equation (2.18) appears valid to $g^2=1$, and the time scales need not be well separated. The series converges in this region to $\Theta = -2i\alpha$, the singularity of $(z-\psi(z))^{-1}$, Equation (2.36), nearest the origin. In this simple model one may solve by inspection Equation (2.18) for Θ obtaining three solutions

$$\Theta = -2i\alpha \qquad g^2 \geq 0$$

$$\Theta = -i\gamma \qquad g^2 > 0$$

$$\Theta = -2i\beta \qquad g^2 > \tfrac{3}{4} \quad .$$

The first is the natural choice for $g^2<1$, the one obtained by iteration, Equation (2.40). The choice of $\Theta = -i\gamma$ in the strong coupling region seems natural for by comparison with the solution, Equation (2.39), it gives in the strong coupling regime the envelope of the oscillations. In true many-body systems no solutions are known for Θ in strong coupling and one has to rely on Equation (2.40).

Let us turn back to general results. It is well known that the weak coupling approximation to Equation (2.18) gives from Equation (2.17) the Pauli equation [39]*

$$\langle\alpha\alpha|\dot{\tilde{\rho}}(t)\rangle = 2\pi\lambda^2\sum_\beta|\langle\alpha|H'|\beta\rangle|^2\delta(E(\alpha) - E(\beta))[\langle\beta\beta|\tilde{\rho}(t)\rangle - \langle\alpha\alpha|\tilde{\rho}(t)]. \quad (2.41)$$

In the Friedrichs model Equation (2.38) is a solution for $g^2 \ll 1$. There $g=\lambda$. This famous result, of course, has the obvious meaning of a gain-loss equation with the transition rates calculated by the "Golden Rule". Equation (2.15) emphasizes that the Pauli equation is a time-asymptotic kinetic equation. Van Hove first remarked that Pauli's original argument (and the numerous later repitions) of repeated random phase approximation was inconsistent [36]. He showed that the argument only

---

*One looks for the lowest order contribution to Θ which is given by the $(\lambda H')^2$ term in $\psi(+ i0)$. We will not repeat this familiar reduction. See reference [13] or [14] for details.

holds at equilibrium. He gave the first good asymptotic derivation by picking from the solution of the von Neumann equation those terms for which $(\lambda^2 t)^n$ is finite as $\lambda \to 0$, $t \to \infty$ [40]. This solution obeys Equation (2.15) in the weak coupling approximation. We should mention that an H-theorem may be easily obtained [5]. By defining

$$H = \sum_{\alpha} \langle\alpha\alpha|\rho(t)\rangle \mathrm{Ln} \langle\alpha\alpha|\rho(t)\rangle \tag{2.42}$$

it may be shown that

$$\dot{H} \leq 0, \tag{2.43}$$

the entropy production ($\dot{S} = -k\dot{H}$)* is positive [41]. In addition it may be seen that the long time solution is

$$\langle\alpha\alpha|\rho(\infty)\rangle = \exp(-\beta H^{o}(\alpha)) \ . \tag{2.44}$$

Equilibrium is determined by the unperturbed energy spectrum in this approximation. The Pauli equation describes asymptotically the approach of a system to equilibrium and from it one obtains the second law of thermodynamics. In the sense of the introduction it describes irreversibility, although only in weak coupling. We must emphasize that the practical applications of this equation are enormous in scope and it is probably the most useful equation in quantum non-equilibrium statistical mechanics. Both the Peierl's equation [11] and the equations of optical pumping [12] are obtaining from it.

In this review we have confined ourselves to closed systems, those isolated from the surroundings. Open systems are difficult to discuss since one must characterize the form of the interaction which may introduce ad-hoc stochastic assumptions analogous to repeated random phase approximations for closed systems [42]. Recently [43] Argyres has

---

*k is of course Boltzmann's constant.

used a projection operator for open systems

$$P^1 = \rho_R(0)\mathrm{Tr}_R \tag{2.45}$$

where $\rho_R(0)$ is the density operator at $t = 0$ for the reservoir and Tr_R is the trace over the reservoir subspace.* It is also assumed initially that $\rho(0) = \rho_s(0)\rho_R(0)$. It is seen that P^1 has the properties of a projector, Equation (2.3), however it is not in general Hermitian nor does it commute with L^0. Immediately one may obtain a G.M.E. for the system. However, the kernel of this equation is difficult to analyze since G(t) contains terms in the reservoir correlation vacuum. In the weak coupling limit where this does not play a role one obtains the Pauli equation for the system

$$\begin{aligned} \langle nn|\dot{\tilde{\rho}}_s(t)) = {} & 2\pi\sum_m |\langle n|H'_s|m\rangle|^2 \delta(E(n) - E(m))[\langle mm|\tilde{\rho}_s(t)\rangle - \langle nn|\tilde{\rho}_s(t)\rangle] \\ & + 2\pi \sum_{m\alpha\beta} |\langle\alpha m|H'_{Rs}|\beta n\rangle|^2 \delta(E(n) + E(\alpha) - E(m) - E(\beta)) \\ & \cdot [\langle\beta\beta|\rho_R(0)\rangle\langle mm|\tilde{\rho}_s(t)\rangle - \langle\alpha\alpha|\rho_R(0)\rangle\langle nn|\tilde{\rho}_s(t)\rangle] . \end{aligned} \tag{2.46}$$

Here H'_s is the internal system interaction, and H'_{Rs} the system-reservoir interaction; $|m\rangle$ and $|\alpha\rangle$ are system and reservoir unperturbed states respectively. Peier and Thellung [44] have used this equation to verify the principle of minimum entropy production [41] valid for an open system in a steady state near equilibrium.

---

*A comprehensive review of open system G.M.E.'s is given by F. Haake in Tracts in Modern Physics 66 ed. by G. Höhler, Springer-Verlag (Berlin, 1973).

CHAPTER III. SUBDYNAMICS [17,18,19,45] - THE $\tilde{\pi}$ PROJECTION

In this section we will demonstrate the existence of an idempotent operator,* $\tilde{\pi}$, which projects from the density operator the asymptotic $\tilde{\rho}_o(t)$, $\tilde{\rho}_o(t) = P\tilde{\pi}\rho(t)$ being governed by Equation (2.17). Although a Hilbert space here strictly speaking does not exist we may along with Prigogine [18] adopt a geometric language and speak of this as a projection onto a "thermodynamic subspace" in the sense that all the thermodynamic equilibrium properties are obtained from $\tilde{\rho}_o(t\to+\infty)$.

Let us first obtain some necessary additional operator relations. Consider the expectation value, $\langle A\rangle_t$, of the observable A

$$\langle A\rangle_t = (A,\exp(-iLt)\rho(0)) = (\exp(+iLt)A,\rho(0)) = (A(t),\rho(0)) \quad , \quad (3.1)$$

where A(t) obeys the Heisenberg equation of motion

$$i\dot{A}(t) = -LA(t) \quad . \quad (3.2)$$

The Hermitian character of L, Equation (1.5), has here led to the equivalence of the Heisenberg and Schrödinger "pictures".

Asymptotically we may write

$$(A,\tilde{\rho}(t)) = (\exp(i\Theta^{\dagger}t)A,\rho(0)) \quad . \quad (3.3)$$

where $\Theta^{\dagger}$ is the Hermitian conjugate of the asymptotic collision operator, Θ Equation (2.18).

Let us introduce the generalization of the notion of velocity inversion [18]

$$L\to-L \equiv L' \quad . \quad (3.4)$$

This with the simultaneous time inversion

$$t\to-t \quad (3.5)$$

---

*This is $\overset{(o)}{\pi}$, Equation (2.12) of GI.

leads to the "time reversal invariance" of the von Neumann equation. We may calculate the inverse evolution with L'

$$(A,\exp(-iL't)\rho(0)) = (\exp(+iL't)A,\rho(0)) \qquad (3.6)$$

and asymptotically

$$(A,\exp(-i\Theta't)\rho(0)) = (\exp(i\Theta'^{\dagger}t)A,\rho(0)) = (\exp(-i\eta t)A,\rho(0)) \qquad (3.7)$$

where

$$\eta = -\Theta'^{\dagger}$$

and

$$\Theta' = \Theta(-L) \ . \qquad (3.8)$$

We may equally well separate out the diagonal part of A(t), with P, Equation (2.1)

$$A(t) = PA + (1-P)A = A_o + A_c . \qquad (3.9)$$

Then, we may in the Heisenberg picture write a generalized master equation for A_o analogous to Equation (2.7) both for the evolution governed by iL or inverse iL'. We may also take the asymptotic limit as in Equations (2.15) and (2.16) and obtain

$$\eta = L_{oo} - i\int_o^{\infty} d\tau \ \exp(i\eta\tau)G(\tau) \qquad (3.10)$$

analogous to Equation (2.18). This may be written as

$$\eta = L_{oo} + DL_{co} \qquad (3.11)$$

where

$$iD = \int_o^{\infty} d\tau \ \exp(i\eta\tau)L_{oc}\exp(-iL_{cc}\tau) \quad . \qquad (3.12)$$

Claude George (see GI, Equation (2.9)) has defined the simultaneous operation of Hermitian conjugation and L inversion as "star Hermitian" conjugation. For instance the generator of the time evolution, iL, is star Hermitian, $(iL)^* = (iL)'^{\dagger} = iL$. Alternatively, we obtain these results by simply making the star Hermitian conjugation of Equations (2.18) and (2.20) and defining

$$\begin{aligned} D &= C^* \\ \eta &= -\Theta^* \ . \end{aligned} \tag{3.13}$$

As can be seen from (2.20) C creates correlations from the vacuum of correlations and from Equation (3.12) D destroys correlations and has been called the destruction operator. C and D are "velocity inverse" Hermitian conjugates.* We should remark in reference [45] as well as [19] the alternative conjugation $\overline{A}$, simultaneous time reversal and Hermitian conjugation has been used. We have here adopted the preferable point of view of reference [18].

Two key relationships are now obtained. We partially integrate Equation (2.20) for Θ and Equation (3.10) for η obtaining

$$C\Theta = L_{co} + L_{cc}C \tag{3.14}$$

$$\eta D = L_{oc} + DL_{cc} \ , \tag{3.15}$$

where the integrated part is assumed to vanish at $\tau = \infty$. One may be obtained from the other by star Hermitian conjugation.

With these operator relations so defined we now proceed to the main point. From Equations (3.14) and (3.15) we write

---

*Equation (2.17) of GI.

$$(P+C)\Theta = L(P+C) \tag{3.16}$$

$$\eta(P+D) = (P+D)L \ . \tag{3.17}$$

Let us define*

$$\tilde{\pi} = (P+C)(1+DC)^{-1}(P+D) \ . \tag{3.18}$$

Immediately from Equation (3.17) we have

$$\tilde{\pi}L = L\tilde{\pi} \tag{3.19}$$

and from

$$(P+D)(P+C) = P(1+DC) = (1+DC)P \tag{3.20}$$

we obtain

$$\tilde{\pi}^2 = \tilde{\pi} \ . \tag{3.21}$$

The operator $\tilde{\pi}$ is idempotent and commutes with L.

Let us separate

$$\rho(t) = \tilde{\pi}\rho(t) + (1-\tilde{\pi})\rho(t) \equiv \tilde{\rho}(t) + \hat{\rho}(t). \tag{3.22}$$

From the formal solution to the von Neumann equation using Equations (3.16), (3.18) and (3.19) we see

$$\tilde{\rho}_o(t) \equiv P\tilde{\pi}\exp(-iLt)\rho(0) = P\exp(-iLt)\,\tilde{\pi}\,\rho(0) = \exp(-i\Theta t)\tilde{\rho}_o(0) \ , \tag{3.23}$$

where

$$\tilde{\rho}_o(0) = P(1+DC)^{-1}(P+D)\rho(0) \ . \tag{3.24}$$

From Equation (3.18)

$$\tilde{\rho}_c(0) = C\tilde{\rho}(0) \ , \tag{3.25}$$

---

*This is Equation (2.15) of GI.

and again using the formal solution to the von Neumann equation

$$\tilde{\rho}_c(t) = (1-P)\exp(-iLt)(P+C)\tilde{\rho}_o(0) = C\exp(-i\Theta t)\tilde{\rho}_o(0) = C\tilde{\rho}_o(t). \quad (3.26)$$

We have the remarkable results which is the main result of this section; the vacuum part of $\tilde{\pi}\rho(t)$, $\tilde{\rho}_o(t)$ and the correlation part $\tilde{\rho}_c(t)$ obey the same equations as in the asymptotic limit discussed in II, Equations (2.17) and (2.19). Here, however, the relationships are valid <u>at all times</u> since Equation (3.23) utilizes the solution to the von Neumann equation. The same notation for $\tilde{\rho}_o(t)$ and $\tilde{\rho}_c(t)$ is here used. However, in section II the equations only appeared to be valid at $t \approx \tau_R$. Differentiating Equations (3.23) and (3.26) we have

$$\begin{aligned} i\dot{\tilde{\rho}}_o &= \Theta\tilde{\rho}_o \\ i\dot{\tilde{\rho}}_c &= C\Theta\tilde{\rho}_o . \end{aligned} \quad (3.27)$$

Equations (3.27) are the <u>generalized kinetic equations</u> for the $\tilde{\rho}_o(t)$ and $\tilde{\rho}_c(t)$ parts of the density operator, valid at all times and uncoupled from $(1-\tilde{\pi})\rho(t)$. Equations (2.17) and (2.19) of chapter II are formal solutions. In this sense they obey an exact and separate dynamics, a subdynamics [17,45]. This follows from the essential property of $\tilde{\pi}$ that it commutes with L, Equation (3.19), in contrast to the P projector introduced in the G.M.E.. The generalized kinetic equations appear in a new light as central to the general theory. They appear markoffian, however, governed by Θ which is a solution to the non-linear operator integral Equation (2.18).

It should be emphasized that the total $\rho(t)$ certainly depends upon $\hat{\rho}(t)$ and this part for reasonable initial condition can be expected to vanish for $t \approx \tau_R$ by a time scaling argument, just as discussed in the previous section. We do not have space to go into details but we have

$$\hat{\rho}(t) = \exp(-iLt)\hat{\rho}(0)$$

and then one may show that $\hat{\rho}_o(t)$ is related to $\hat{\rho}_c(t)$,

$$\hat{\rho}_o(t) = -D\hat{\rho}_c(t) ,$$

and $\hat{\rho}_c(t)$ obeys a separate general kinetic type equation.

Before outlining some further results of this separation we should remark that important features are not apparent in this heuristic operator manipulation. These are perhaps more transparent in the perturbation theory development. The general kinetic equation must exist. This condition has already been stated in the first relationship of Equation (2.14) and its equivalent in Laplace transforms discussed there. The Friedrichs model illustrates this point, and we do not expect this to hold for instance for long range potential like the gravitational interaction. The thermodynamic limit plays also an essential role technically leading to the continuous spectra and Cauchy integrals in matrix elements of Equation (2.14) such as mentioned in connection with the Friedrichs model. In addition Equation (2.14) must not vanish which was stated as the dissipative condition of Equation (2.34). Physically, there must be collisions in the system.

Finally we should mention that the subdynamics separation has been formulated in reduced distribution functions by Balescu [46]. We have adopted here the N-body operator formulation for conciseness of presentation. We should also mention the mathematical discussions of the group at Milan [47] and also Bongoarts, Fannes, and Verbeure [48], the latter being in the thermodynamic limit utilizing C* algebras. The results mentioned above have recently been verified in the Friedrich's model by De Hann and Henin [49]. There however, the primary aim was to investigate the causal representation, the star unitary transformation discussed in GI.

Let us now turn to some results of the $\tilde{\pi}$ separation. First consider invariants of the motion [50]. Let I be an invariant,

$$LI = 0 \ . \tag{3.28}$$

We obtain easily in the manner of Equation (2.5)

$$L_{oo}I_o + L_{oc}I_c = 0 \tag{3.29a}$$

$$L_{co}I_o + L_{cc}I_c = 0 \ . \tag{3.29b}$$

Multiply by $-i\ \exp(-iL_{cc}t)$ and integrate

$$-i\int_o^t d\tau\{\exp(-iL_{cc}\tau)L_{co}I_o + [\exp(-iL_{cc}t) - 1]I_c = 0 \ . \tag{3.30}$$

In the limit $t \to \tau_R$ applying the first condition, Equation (2.14) we obtain

$$I_c = -i\int_o^\infty d\tau\ \exp(-iL_{cc}\tau)L_{co}I_o \tag{3.31}$$

where the limit $t \simeq \tau_R \to \infty$ since the integrand does not contribute in this range. We have

$$[L_{oo} - iL_{oc}\int_o^\infty d\tau\ \exp(-iL_{cc}\tau)L_{co}]I = 0 \ . \tag{3.32}$$

By comparison with Equations (2.20) and (2.21) we have

$$[L_{oo} + L_{oc}C]I = \Theta I = 0 \ , \tag{3.33}$$

and from Equation (3.20)

$$\tilde{\pi}I = \tilde{\pi}(P+C)I_o = \pi \ . \tag{3.34}$$

Thus

$$\tilde{\pi}I = 0 \ , \tag{3.35}$$

and invariants of the motion are contained solely in the $\tilde{\pi}$ projection. The main point we want to draw from Equation (3.33) is that $\rho(H)$ is an invariant and thus contained in the $\tilde{\pi}\rho$ part of the density operator. It is the time independent solution to the general kinetic equation and governs the thermodynamic behaviour of the system.

If we use Equation (3.29) in Equation (3.30) we obtain

$$[L_{oo} - iL_{oc}\int_o^t d\tau \exp(-iL_{cc}\tau)L_{co}]I_o + L_{oc}\exp(-iL_{cc}t)I_c = 0 \ . \qquad (3.36)$$

This result has provided the natural classification into regular and singular invariants, the former those for which $t \approx \tau_R$ the last term vanishes, and we may put $t\to\infty$ in the limit exactly as discussed. Regular invariants are the zero eigenvalues of the collision operator Θ, Equation (3.33), or what is necessary,

$$\psi(+io)I_o = 0 \ . \qquad (3.37)$$

In their analysis of the Friedrichs model discussed in chapter II, Grecos and Prigogine have considered the invariants [38]. There they construct the exact eigenvalues for a general Friedrichs model, and then the I are just linear superposition of density matrices constructed from these states. By examining the second term in Equation (3.36) they show that these invariants are singular. In addition it is shown that these invariants cannot be expressed in terms of a perturbation expansion analytic in the coupling constant with the unperturbed state, $|E\rangle$, $|\omega\rangle$, of Equation (2.23).

Finally, let us close this section by considering relativistic covariance [51]. Based on the work of Dirac [52], Balescu and Kotera have formulated a classical relativistic statistical mechanics [53]. The idea is that all transformations of the Poincaré group can be represented by canonical transformations in phase space. The generators are the Hamiltonian H, generating time translation; the three generators of spatial rotation, R; and the generators of the transformations from one reference frame to another moving with constant velocity, K, the Lorentz transformation.* Relativistic covariance of ρ is assured by solving a set of Liouville-like equations. One of the generalized "Liouvillians" is precisely L_H defined by Equation (1.5). They satisfy the commutation properties of the group,

$$
\begin{aligned}
[L_H, L_R]_- &= 0 \\
[L_K, L_H]_- &= L_R \\
[L_K, L_R]_- &= L_H \quad .
\end{aligned}
\tag{3.38}
$$

Here, of course, L_H, L_K, and L_R are appropriate poisson brackets (commutators) with the H, R and K mentioned earlier. From this it has been proved that the Liouville equation is Lorentz invariant [53]. Two observers in different frames describe the evolution by the same Liouville equation having the same generator L_H but with differing initial conditions. The time appearing in each equation is the time in the respective rest frame. Since the full G.M.E. (for ρ_o and ρ_c) is equivalent to the Liouville equation it is expected that it too should be Lorentz invariant. This has been verified [53].**

The important question is, "is the concept of subdynamics relativistic invariant?" This has been answered by Balescu and Brenig [51].

It is easy to prove

$$
\begin{aligned}
[L_H, \tilde{\pi}]_- &= 0 \\
[L_R, \tilde{\pi}]_- &= 0 \quad .
\end{aligned}
\tag{3.39}
$$

The former has already been proven in Equation (3.19) and the latter

---

*Along with Balescu et al, we will here actually consider a subgroup of the Poincaré group.

**A word should be said again about notation. In Balescu's papers, our $L \to \mathcal{L}$, $P \to V$, $(I-P) \to C$, $\tilde{\pi} \to \pi$, and $\Theta_H = V\Gamma_H V$ and $\psi(z) \to \mathcal{E}(z)$. This notation is used in reference [16].

follows from the commutation of L_R with L_H and P. P is here taken to satisfy

$$[L_H^o, P]_- = 0$$

$$[L_R^o, P]_- = 0 \tag{3.40}$$

$$[L_K^o, P]_- = 0 ,$$

the first of these being obeyed by the realizations of P given in Equation (2.1). These express the physical idea that correlations are not affected by the "unperturbed motion". It must be stressed that P does not commute with L_H or L_K. Thus the separation of ρ into ρ_o and ρ_c is not an invariant separation under a Lorentz transformation.

The proof of

$$[L_K, \tilde{\pi}]_- = 0 \tag{3.41}$$

is unfortunately not algebraically simple. Most important it is necessary to again assume the condition that certain operator combinations of $\psi(z)$, and L_K are regular in the neighborhood of $z = 0$ exactly as discussed earlier. Certainly this is the condition as Equation (2.14) with an enlarged class of A's involving L_K. L_K, it must be realized, contains terms involving the interparticle potential (interacting electromagnetic field) just as L_H does. It is not at all surprising that such a condition should appear.

With Equation (3.41) the Lorentz invariance of $\tilde{\pi}$ and $\hat{\pi}$ are established. This means that $\tilde{\rho}$ and $\hat{\rho}$ are invariant manifolds. Under a Lorentz transformation the $\tilde{\rho}$ part transforms into a $\tilde{\rho}$. As might be expected they obey the same set of general kinetic-like equations of the form of Equations (3.27) for all the elements H, R, K with an Θ_H, Θ_R, and Θ_K generating the evolution and transformation properties. The transformation laws for $\tilde{\rho}_o$ between Lorentz frames is governed by a kinetic-like equation containing the generator $\Theta_H - t\Theta_R$. The relationship between $\tilde{\rho}_o$ and $\tilde{\rho}_c$, Equation (2.19), is the same in all Lorentz frames. Similar operators are constructed for $\hat{\rho}_c$. Then it may be shown that these generators obey a Lie algebra of the form Equation (3.38). Thus, the relativistic covariance of the generalized kinetic equation is assured. The general kinetic equation, for $\tilde{\rho}_o$, Equation (2.17) is <u>form invariant</u> under a Lorentz transformation. As with the Liouville equation only the initial conditions differ between Lorentz frames.

The deep and consistent nature of the separation of ρ into $\tilde{\rho}$ and $\hat{\rho}$ is now apparent from this last result. Generalized kinetics as described by the relativistic covariant equation for $\tilde{\rho}_o$ governed by Θ_H appears as a reduced dynamics. It is a separate dynamics governing the irreversible kinetic behavior of many body systems. The representation of the dynamical behavior of many body systems in the infinite limit is in a sense of group theory reducible. This is a surprising and unlooked for result, even if it is to be understood in a weak sense.

CHAPTER IV. QUANTUM KINETICS OF DILUTE SYSTEMS

For the dilute atomic or molecular gas as mentioned in the Introduction the characteristic observables are hydrodynamic quantities such as the average one particle density, velocity, and kinetic energy and the transport coefficients. In this system correlations are not readily observed. It is our main object here to obtain the quantum Boltzmann equation [21,55,56,57,58] and to show explicitely the relation with the concepts of scattering theory. Proceeding from the G.M.E. of section II in the case of spatially homogenous systems one may introduce a realization of P in which the $|\alpha\rangle$, Equation (2.1), are the product plane wave states of N non-interacting particles. One may also consider the solution of the G.M.E. in the plane wave representation and by diagrammatic methods obtain those terms which are lowest order in the density [59]. One still must reduce these results to kinetic equations involving single particle distribution functions which involves the justification of the factorization of two-body distributions (molecular chaos problem).

The realization of an appropriate P for spatially inhomogeneous systems in a consistent manner has proved difficult [60] and is now only being resolved with the work of Balescu on the dynamical correlation patterns [46]. For our purposes here it is still most convenient when discussing the dilute quantum gas to form from the von Neumann equation a coupled set of equations for reduced distribution functions, the B.B.G.K.Y. hierarchy [21,61,62,63,64]. From this we will concisely obtain the quantum version of the Boltzmann equation [20] and make a connection to formal binary scattering theory.

In the following we will confine ourselves to Boltzmann statistics; the proper symmetry may be introduced directly [20]. The principle quantum effect is here the diffraction effect which is important when the average de Broglie wave length is of the order of the interparticle distance,

$$\hbar(2mkT)^{-\frac{1}{2}}c^{\frac{1}{3}} \approx 1$$

We will not have space to here discuss the most interesting interference effects due to degeneracy of the internal states of the molecules, which lead to such "exotic" effects as the Senftleben-Beenakker effect, alignment of non-spherical molecules by transport processes, etc. [65, 66]. These are described by a generalization of the Boltzmann equation due to Waldmann [67] and Snider [57]. It has been obtained from the G.M.E. giving a consistent characterization of the composite particles by Grecos and Schieve [68]. Closely related to quantum kinetics is the theory of collision broadening of the spectral line [69,70]. This has been discussed from a point of view close to that presented here by Roney [71].

The most important physical application of the quantum Boltzmann equation is the calculation of the quantum corrections to the transport coefficients by the method of Chapman and Enskog [72,73,74]. This method of solution is an expansion in $\frac{\tau_R}{\tau_H}$, the ratio of the relaxation time to the hydrodynamic time. The quantum modification of the classical solution is quite straight forward incorporating quantum scattering phase shifts in expressions for the transport coefficients.

We will adopt a concise analysis of the hierarchy due to Tip [75] and Hawker [76]. Other methods are possible such as the quantum version [77] of the analysis of the hierarchy of E.G.D. Cohen [78]. We assume the N-body Hamiltonian may be written as

$$H(1,2,\ldots N) = \sum_{i=1}^{N} T(i) + \sum_{i>j=1}^{N} H'(ij) , \qquad (4.1)$$

where the single particle kinetic energies are $T(i)$ and $H'(ij)$ is the two particle interaction potential. The density operators $\rho(1,2\ldots N,t)$ obeys a von Neumann equation in the following form

$$i\dot{\rho}(1\ldots N,t) = [\sum_{i=1}^{N} T(i) + \sum_{i>j=1}^{N} H'(ij),\rho(1,2\ldots N,t)] \ . \qquad (4.2)$$

Define reduced density operators by

$$\rho(1,2\ldots s,t) = V^{s} \operatorname*{Tr}_{(s+1,\ldots N)} \rho(1,2\ldots N,t) \ . \qquad (4.3)$$

Here V is the volume of the system and the partial trace is taken over the set (s+1,...N). We multiply Equation (4.2) by V and take the partial trace, $\operatorname*{Tr}_{2\ldots N}$. Then, we use the cyclic trace property and the vanishing of the partial trace of such commutators,

$$\operatorname*{Tr}_{2\ldots N}[T(i\neq 1),\rho(1\ldots N,t)] = 0 \ ,$$

obtaining

$$i\dot{\rho}(1,t) = [T(1),\rho(1,t)] + V \sum_{i>j=1}^{N} \operatorname*{Tr}_{(2\ldots N)} [H'(ij),\rho(1\ldots N,t)].$$

Further, assuming the symmetry of $\rho(1\ldots N,t)$ under exchange of particle labels (identical particles) and taking the thermodynamic limit,

$$\lim_{\substack{N\to\infty \\ V\to\infty}} \frac{(N-1)}{V} = \frac{N}{V} \equiv c = \text{constant}$$

we have the first equation of the hierarchy

$$i\dot{\rho}(1,t) = L(1)\rho(1,t) + c\operatorname*{Tr}_{2} L'(12)\rho(12,t) \ . \qquad (4.4)$$

where

$$L(1) = [T(1),]$$

$$L(12) = [T(1) + T(2) + H'(12),]$$

$$= L^{o}(12) + L'(12) \ .$$

etc.

In the same manner

$$i\dot{\rho}(12,t) = L(12)\rho(12,t) + c \operatorname*{Tr}_{3}\{(L'(13) + L'(23))\rho(123,t)\}. \quad (4.5)$$

Subsequent elements of hierarchy may be obtained, the last element for finite systems being the von Neumann equation itself. To obtain the Boltzmann equation we shall only need these first two elements.

Define the two and three body correlation operators

$$g(12) = \rho(1)\rho(2) - \rho(12)$$

$$\begin{aligned} g(123) = \rho(123) - \rho(1)g(23) - \rho(2)g(13) \\ - \rho(3)g(12) - \rho(1)\rho(2)\rho(3) . \end{aligned} \quad (4.6)$$

Here, for instance, if $g(12) = 0$ then $\rho(12)$ is uncorrelated, factoring into $\rho(1)\rho(2)$. In the momentum representation for spatially homogenous systems this definition of correlations may be shown to be the same as that defined as $\rho_c = (1-P)\rho$ with the realization of P discussed at the beginning of this section.

Now, using the definition of g(12) in the second equation of the hierarchy, Equation (4.5), we obtain an equation involving $\partial_t[\rho(1,t)\rho(2,t)]$. This may be rewritten using the first equation, Equation (4.4) and we obtain from Equation (4.5)

$$\begin{aligned} i\dot{g}(12,t) = L(12)g(12,t) + L'(12)\rho(1,t)\rho(2,t) \\ + c\operatorname*{Tr}_{3}\{L'(13)[\rho(1,t)g(23,t) + \rho(3,t)g(12,t) + g(123,t)] \\ + L'(23)[\rho(2,t)g(13,t) + \rho(3,t)g(12,t) + g(123,t)]\} . \end{aligned} \quad (4.7)$$

To the lowest order in the density we then have

$$i\dot{g}(12,t) = L'(12)\rho(1,t)\rho(2,t) + L(12)g(12,t) + O(c). \quad (4.8)$$

We may formally solve this, just as with the G.M.E., Equation (2.5b), in section II, taking the causal particular solution (see comments on page 10)

$$ig(1,2,t) = \int_0^t d\tau \exp(-iL(12)\tau)L'_{12}\rho(1,t-\tau)\rho(2,t-\tau)$$

$$+ \exp(-iL(12)t)g(12,0) \qquad (4.9)$$

$$+ O(c^2) .$$

Using this result in the $c\mathrm{Tr}_2$ term of Equation (4.4) we have

$$i\dot{\rho}(1,t) = L^o(1)\rho(1,t) + c\mathrm{Tr}_2 L'(12)\rho(1,t)\rho(2,t)$$

$$+ c\mathrm{Tr}_2 L'(12) \exp(-iL(12)t)g(12,0) \qquad (4.10)$$

$$+ (-i)c\mathrm{Tr}_2\int_0^t d\tau\ L'_{12}\exp(-iL(12)\tau)L'(12)\rho(1,t-\tau)\rho(2,t-\tau)$$

$$+ O(c^2) .$$

This equation has the same structure as the G.M.E.; however, being interested in terms to the lowest order in c we must expand

$$\rho(1,t-\tau) = \rho(1,t) - \tau\dot{\rho}(1,t)...$$

and iterate in the non-Markoffian term. Keeping <u>all orders in τ</u> and lowest order in c we have under the integral

$$\rho(1,t-\tau)\rho(2,t-\tau) = \exp(iL^o(12)\tau)\rho(1,t)\rho(2,t) + O(c). \qquad (4.11)$$

Now using

$$\exp(-iL(12)\tau)L'(12) \exp(iL^o(12)\tau) = id_\tau[\exp(-iL(12)\tau) \exp(iL^o(12)\tau)];$$

the τ integral contains a perfect differential. The contribution at $\tau=0$ exactly cancels the second term on the right hand side of Equation (4.10). We have then

$$i\dot{\rho}(12,t) = L^o(1)\rho(1,t) + c\mathop{Tr}_2 \exp(-iL(12)t)\exp(iL^o(12)t)\rho(1,t)\rho(2,t)$$

$$+ c\mathop{Tr}_2 L'(12)\exp(-iL(12)t)g(12,0) + O(c^2)\ .$$

This result is valid at all times to this order in c. We now make the asymptotic approximation.

$$\exp(-iL(12)t)\exp(iL^o(12)t)\rho(1,t)\rho(2,t)$$

$$\approx \lim_{t'\to-\infty} \exp(-iL(12)(t-t'))\exp(+iL^o(12)(t-t'))\rho(1,t)\rho(2,t) \tag{4.12}$$

$$= \Omega^+(12)\rho(1,t)\rho(2,t)(\Omega^+(12))^\dagger .$$

This is valid for $t \approx \tau_R >> \tau_c$, the collision duration time $\tau_c = \frac{\sigma}{<v>}$, σ being the interaction range. The collision occurs at t' = 0 and the $\exp(-iL(12)t)\exp(iL^o(12)t)$ is unity outside the interaction sphere.

$\Omega^+(12)$ is the outgoing Moller wave operator of formal scattering theory [6,79].

$$\Omega^+(12) = \lim_{\tau\to\infty} \exp(-iH(12)\tau)\exp(iH^o(12)\tau) \tag{4.13}$$

having the properties

$$H\Omega^+ = \Omega^+ H^o$$

$$(\Omega^+)^\dagger\Omega^+ = I \tag{4.14}$$

$$\Omega^+(\Omega^+)^\dagger = I - \Lambda$$

where Λ is the projection operator on the sub-space of two particle bound states. From this it may be seen that Ω^+ transforms the continuous spectrum of $H^o = T(1) + T(2)$ into that of H(12). It is related to the scattering matrix, the T^+ matrix, by*

---

*These relationships are trictly true only for half-off T matrices, for instance

$$<\alpha|T^+(E_\beta)|\beta> = <\alpha|H'\Omega^+|\beta>.$$

$$T^+ = H'\Omega^+$$
$$\Omega^+ = I + G^o(+io)T^+ \tag{4.15}$$

where

$$G^o(+io) = \lim_{\varepsilon\to o^+}(i\varepsilon - H^o)^{-1} .$$

The T^+ matrix is related to the transition rate between states $|p\rangle$ and $|p'\rangle$ of the unperturbed Hamiltonian by the well known rule

$$\omega(p'\leftarrow p) = 2\pi|\langle p'|T|p\rangle|^2\delta(E(p') - E(p)) . \tag{4.16}$$

In the following we will for convenience drop the + notation on the outgoing Möller wave operator. With the asymptotic approximation we have

$$i\dot{\rho}(1,t) = L_1^o\rho(1,t) + c\,\underset{2}{\mathrm{Tr}}L'(12)\Omega(12)\rho(1,t)\rho(2,t)\Omega^\dagger(12)$$
$$+ c\,\underset{2}{\mathrm{Tr}}L'(12)\exp(-iL(12)t)g(12,0) + O(c^2) . \tag{4.17}$$

Consistent with the asymptotic approximation above, the final assumption is either to take initially $g(12,0) = 0$ or to argue that on some time scale $t\to\tau_R$ that the initial correlations will decay just as was discussed with respect to the G.M.E. of section I. With this we obtain a closed operator equation for $\rho(1,t)$ to lowest order in the density, c,

$$i\dot{\rho}(1,t) = L^o(1)\rho(1,t) + c\,\underset{2}{\mathrm{Tr}}L'(12)\Omega(12)\rho(1,t)\rho(2,t)\Omega^\dagger(12). \tag{4.18}$$

This is a generalization of the Boltzmann equation. As we shall see shortly it contains spatial delocalization effects as well as effects of incomplete collisions since the T-matrices are off-shell. We note the "factorization" of $\rho(12,t)$; $g(12,t)$ does not appear. This is an operator version of molecular chaos, which has resulted from the density expansion and the neglect of g(12,0).

To explicitly see the spatial dependence it is convenient to introduce the Wigner function [80,81,82]

$$f(\underset{\sim}{r},\underset{\sim}{p},t) = (2\pi)^{-3}\int dk\, \exp(i\underset{\sim}{k}\cdot\underset{\sim}{r})\langle \underset{\sim}{p} + \frac{\underset{\sim}{k}}{2}|\rho(1,t)|\underset{\sim}{p} - \frac{\underset{\sim}{k}}{2}\rangle. \quad (4.19)$$

It is not truly a distribution function since f may be negative. However, average values are well defined. It may be shown [81,82] that

$$\langle A\rangle_t = \int d\underset{\sim}{p}d\underset{\sim}{r}\, A(\underset{\sim}{r},\underset{\sim}{p})f(\underset{\sim}{r},\underset{\sim}{p},t) \quad (4.20)$$

where $A(\underset{\sim}{r},\underset{\sim}{p})$ is the classical dynamical variable corresponding to the quantum observable A. This choice of distribution function assumes the Weyl [83] correspondence rule for classical products $r^m p^n$,

$$r^n p^m \leftrightarrow 2^{-n} \sum_{1=0}^{n} \binom{n}{1} r_{op}^{n-1}\, p_{op}^{n}\, r_{op}^{1}\ . \quad (4.21)$$

Other possible correspondence rules are possible [84] leading to alternative f's. The Weyl rule leads to the simplest phase space distribution function. This point in connection to kinetics theory has been discussed by Hawker [85].

Let us sketch how the quantum Boltzmann equation for f(r,p,t) is obtained from Equation (4.17). Adopt the plane wave representation $|p_1 p_2\rangle$*

$$H^o|p_1p_2\rangle = (T(1) + T(2))|p_1p_2\rangle$$
$$= (\frac{p_1^2}{2m} + \frac{p_2^2}{2m})|p_1p_2\rangle$$

with the normalization

---

*We drop the explicit vector notation on $\underset{\sim}{p}$.

$$\langle p_1|p_2\rangle = \delta(p_1-p_2). \tag{4.22}$$

We have

$$\langle p_1p_2|H'|p_1'p_2'\rangle = \langle\frac{p_1-p_2}{2}|H'|\frac{p_1'-p_2'}{2}\rangle\delta(p_1+p_2-p_1'-p_2')$$

where

$$\langle a|H'|b\rangle = H'(a-b) \tag{4.23}$$

because of conservation of momentum in binary collision and the assumed locality of the scattering potential. Take the matrix elements of Equation (4.18) in this representation, then Fourier transform it with respect to $k = (p_1-p_1')$. Thus, from the definition, Equation (4.19) we obtain

$$i\dot{f}(r,p,t) = -im^{-1}p\cdot\nabla_r f(r,p,t) + \mathcal{B}(r,p,t), \tag{4.24}$$

where $p = \frac{p_1+p_1'}{2}$ and $\mathcal{B}(r,p,t)$ is the collision term. As an illustration consider the free flow term in Equation (4.18)

$$\begin{aligned}&(2\pi)^{-3}\int dk\ \exp(ik\cdot r)\langle p_1|L^o\rho(1,t)|p_1'\rangle\\
&= (2\pi)^{-3}\int dk\ \exp(ik\cdot r)\left(\frac{p_1^2}{2m} - \frac{p_1'^2}{2m}\right)\langle p + \frac{k}{2}|\rho(1,t)|p - \frac{k}{2}\rangle\\
&= -im^{-1}p\cdot\nabla_r f(r,p,t)\ .\end{aligned}$$

We have used

$$\frac{1}{2m}(p_1^2 - p_1'^2)\ \exp(ik\cdot r) = -\frac{i}{m}\,p\cdot\nabla_r\ \exp(ik\cdot r)\ .$$

After some manipulation and the use of relationship Equation (4.23) we obtain

$$B(r,p,t) = (2\pi)^{-3} c \int dq\, dq_1\, dq_2\, dk\, dx\, dy \; \exp(-i(x\cdot q_1 + y\cdot q_2))$$

$$\cdot\; f(r+\frac{x+y}{2}, p-q+k)\, f(r+\frac{x-y}{2}, p-q-k)\{<p+\frac{q_1}{2}|H'\Omega|k+\frac{q_2}{2}><k-\frac{q_2}{2}|\Omega^{\dagger}|q-\frac{q_1}{2}> \qquad (4.25)$$

$$- <q+\frac{q_1}{2}|\Omega|k+\frac{q_2}{2}><k-\frac{q_1}{2}|\Omega^{\dagger}H'|q-\frac{q_1}{2}>\} \;.$$

The non-locality of the collision term and the off energy shell character of the scattering matrix in $B(r,p,t)$ are now apparent. It is spatially and in terms of scattering temporally delocalized. The latter is seen by the fact the collision duration in one definition may be defined as [86] $\mathrm{Re}[-iT^{-1}\,\frac{\partial T}{\partial E}]$.

To localize it we assume

$$f(r+\frac{x+y}{2}) = f(r) + O(\nabla_r) + \ldots$$

and neglect the higher gradient terms. The first order in gradient terms lead to the so-called collisional transfer corrections to the Boltzmann equation [63,85,87,88]. Then we use the relations of scattering theory given by Equations (4.15) and the optical theorem [6]

$$\mathrm{Im}<k|T|k> = -\pi\int dq\,|<k|T|q>|^2\delta(E(k) - E(q))$$

and

$$\lim_{\varepsilon\to 0}(X + i\varepsilon)^{-1} = \frac{P}{X} - i\pi\delta(X).$$

The δ function part only contributes in $B(rp,t)$ and we obtain finally in relative particle coordinates the quantum Boltzmann equation.

$$\dot{f}(rp,t) = -m^{-1}p\cdot\nabla_r f(rp,t) + 8(2\pi)^4 c\int dk\,dq\;\delta(E(k) - E(q))\,|<k|T|q>|^2 \qquad (4.26)$$

$$\{f(r,p-q-k,t)f(r,p-q+k,t) - f(r,p,t)f(r,p+2k,t)\} \;.$$

The kernel is related to the transition rate and the quantum differential cross section by Equation (4.16),

$$\frac{d\sigma}{d\Omega_k}(q \to k) = m^2(2\pi)^4 |\langle k|T|q\rangle|^2 \ .$$

The T's are now on the energy shell because of the conservation law. We note that in contrast to the Pauli equation mentioned in section II it is strong interaction equation valid for large deflections. The T-matrix obeys the well-known Lippman-Schwinger equation of elastic-binary scattering theory [6,79].

$$T(12) = H'(12) + H'(12)G^o(+io)T(12). \qquad (4.27)$$

Like the classical Boltzmann equation it has the gain-loss form and may be derived intuitively including the proper quantum statistics [21] which we have not included.

The hydrodynamic equations (conservation laws of particle number, energy, momentum) may be derived immediately. Also, it may be shown that the local Maxwellian is an equilibrium solution. The local Maxwellian is the starting point (zeroth order) in the Chapman-Enskog theory [72]. Just as in the classical theory an *H*-theorem (entropy principle) may be proved

$$\dot{H} \leq 0 \qquad (4.28)$$

where

$$H(t) = \int dr dp \ f(rp,t) \mathrm{Ln} f(rp,t) \ . \qquad (4.29)$$

In general no reflection-rotation symmetry of the cross section need be assumed to prove this. As emphasized by Stueckelberg only unitarity of the S-matrix* is required [89]. Thus, the theorem is valid for polyatomic molecules. The only generalization of Boltzmann's *H*-theorem in terms of reduced distribution functions is to include internal structure using the Waldman-Snider equation [57,65,75]. However, the

---

* $S = (\Omega^-)^+\Omega^+$ where Ω^- is the incoming Möller operator.

recent results of the star-unitary transformation mentioned in the Introduction (see G-I) imply a general N-body *H*-theorem [18], having the form of the transformed square modulus of $\rho_N(t)$. The connection of this theorem to reduced distribution functions is being investigated.

To the next order in the density, c^2, the analysis of the hierarchy leads to the quantum form of the Choh-Uhlenbeck equation [77,78].* For this equation no *H*-theorem has yet been established. Also to this order in two dimensions divergences have been found classically in the transport coefficients [90,91,92,] (in three dimensions in the c^3 order in the density). These divergences arise from repeated binary collisions (A-B, A-C, A-B). Classically for the hard sphere Lorentz gas model (B and C being random massive scatterers) the cross section for this process is $\sigma(\frac{\sigma}{R})^2$ where σ is the diameter of B and C, and R is their distance apart. Upon randomly averaging over the arrangements of B and C in two dimensions the effective cross section goes as $\lim_{R\to\infty} \operatorname{Ln} R$. On the average the duration of this repeated binary collision is ∞ and the time scaling discussed earlier in connection with the asymptotic approximation is no longer valid. The quantum mechanical random impurity problem has been studied by Neal [94] and Resibois-Velarde [95]. Neal finds after a resummation of diagrams to all orders in c that the electrical resistivity is non-analytic function of the impurity density, c, having the form in three dimensions

$$\text{Resistivity} = \alpha c + \beta c^2 + \gamma c^3 \operatorname{Ln} c + \delta c^3 + \dots .$$

The same results have been found earlier for the viscosity and thermal conductivity of the classical gas. The effects of the characteristic logarithmic term are small compared to other terms and it has not been possible yet to demonstrate its existence from the transport theory

---

*Similar procedures to those employed here give a concise derivation. See reference 85 and K. Hawker and W. C. Schieve in preparation.

data [96]. However, in a computer model of interacting hard disks a long time persistence to the velocity auto-correlation function decaying as $(\frac{\tau}{\tau_R})^{-1}$ have been observed for 25 collisions/particle [97]. The decay seems to have its origin in the correlated sequence of binary collisions discussed above.

Even with these theoretical subtleties we are astounded at the extent to which the classical and quantum Boltzmann equations remain the archetypes of non-equilibrium statistical mechanics. The insight of Boltzmann withstood and was reinforced by the profound changes in our understanding of mechanics that occurred with the advent of quantum mechanics.

ACKNOWLEDGMENTS

I particularly want to thank J. W. Middleton and K. E. Hawker for their many contributions to this article. I also want to thank Professor Prigogine and his group at Brussels, particularly Radu Balescu, for their hospitality during this summer when this was completed. Acknowledgement should be given to NATO under Research Grant Number 644 for making the visit possible.

References:

[1] J. VON NEUMANN: Gott. Nachr., 245, 273 (1927).

[2] L. LANDAU: Z. Physik 45, 430 (1927)

[3] J. VON NEUMANN: Mathematical Foundations of Quantum Mechanics (Princeton University Press, 1955).

[4] U. FANO: Rev. Mod. Phys. 29, 74 (1957).

[5] R. TOLMAN: The Principles of Statistical Mechanics (Oxford University Press, London, 1938), Chapter 9.

[6] R. NEWTON: Scattering Theory of Waves and Particles (McGraw-Hill, New York, 1966), Chapter 8.

[7] L. BOLTZMANN: "Further Studies on the Thermal Equilibrium of Gas Molecules", translated and collected by S. G. Brush, Kinetic Theory Volume II (Pergamon Press, 1966).

[8] E. ZERMELO: "On a Theorem of Dynamics and Mechanical Theory of Heat", ibid. page 208.

[9] I. SEGAL: Mathematical Problems of Relativistic Physics (Am. Math. Soc., Providence, Rhode Island, 1963).

[10] G. EMCH: Algebraic Methods in Statistical Mechanics and Quantum Field Theory (Wiley-Interscience, New York, 1972).

[11] R. E. PEIERLS: Quantum Theory of Solids (Oxford University Press, Oxford, 1955).

[12] W. HAPPER: Rev. Mod. Phys. 44, 169 (1972).

[13] I. PRIGOGINE: Non-Equilibrium Statistical Mechanics (Interscience, New York, 1962).

[14] R. ZWANZIG: Lectures in Theoretical Physics, Volume 3 (Boulder, 1960), 106. See also Physica 30, 1109 (1965).

[15] J. W. MIDDLETON AND W. C. SCHIEVE: Physica 63, 139 (1973).

[16] R. BALESCU: Lectures in Statistical Physics, W. C. Schieve, M. G. Velarde and A. P. Grecos, eds. (Springer-Verlag, New York, 1971), page 149.

[17] I. PRIGOGINE, C. GEORGE AND F. HENIN: Physica 45, 418 (1969).

[18] I. PRIGOGINE, C. GEORGE, F. HENIN AND L. ROSENFELD: "A Unified Formulation of Dynamics and Thermodynamics", Chemica Scripta 4, 5 (1973).

[19] W. C. SCHIEVE AND I. PRIGOGINE: "The Role of Subdynamics in Kinetic Theory", Eighth Int. Symposium on Rarefied Gas Dynamics, proceedings to appear (Academic Press).

[20] E. A. UEHLING AND G. E. UHLENBECK: Phys. Rev. 43, 553 (1933).

[21] N. N. BOGOLIUBOV: Lectures on Quantum Statistics I, (Gordon and Breach, New York, 1967).

[22] G. EMCH: Helv. Phys. Acta 37, 67 (1964).

[23] H. PRIMAS: Helv. Phys. Acta 34, Appendix (1961).

[24] S. NAKAJIMA: Progr. Theor. Phys. 20, 948 (1958).

[25] R. BALESCU: Statistical Mechanics of Charged Particles (Interscience-John Wiley, New York, 1963).

[26] B. LEAF AND W. C. SCHIEVE: Physica 30, 1389 (1964).

[27] I. PRIGOGINE AND P. RÉSIBOIS: Physica 27, 629 (1961).

[28] G. JONES: J. Math. Phys. 5, 653 (1964).

[29] G. C. SEWELL: Lectures in Theoretical Physics, Volume 10 (Boulder, 1967), 289.

[30] N. N. BOGOLIUBOV: Journal Phys. U.S.S.R. 10, 265 (1946).

[31] L. LANZ AND L. A. LUGIATO: Physica 47, 345 (1970).

[32] P. RÉSIBOIS: Physica 27, 541 (1961).

[33] K. O. FRIEDRICHS: Commun. Pure and Applied Math. 1, 361 (1940).

[34] J. L. PEITENPOL: Phys. Rev. 162, 1301 (1967).

[35] I. PRIGOGINE, C. GEORGE AND F. HENIN: Proc. Nat. Acad. Sci. USA 65, 789 (1970).

[36] L. VAN HOVE in E.G.D. Cohen: Fundamental Problems in Statistical Mechanics (North Holland, Amsterdam, 1962).

[37] K. HAUBOLD: Physica 28, 834 (1962).

[38] A. P. GRECOS AND I. PRIGOGINE: Physica 59, 77 (1972).

[39] W. PAULI: Festschr. Zum 60, Geburstag A. Sommefeld (Hirzl, Leipzig, 1928).

[40] L. VAN HOVE: Physica 21, 517 (1955).

[41] I. PRIGOGINE: Thermodynamics of Irreversible Processes (Interscience, New York, 1955).

[42] P. G. BERGMANN AND J. L. LEBOWITZ: Phys. Rev. 99, 59 (1955).

[43] P. N. ARGYRES: Lectures in Theoretical Physics, Volume 8A (Boulder, 1966), 183. See also W. PEIER: Physica 57, 565 (1972).

[44] W. PEIER AND A. THELLUNG: Physica 46, 577 (1970).

[45] C. GEORGE, I. PRIGOGINE AND L. ROSENFELD: Mat. Fys, Medd. Dan. Vid. Selsk. 38 (1972).

[46] R. BALESCU: "Non-Equilibrium Statistical Mechanics", Irreversibility in Many-Body Problems ed. by J. Biel and J. Rae (Plenum, New York, 1972); also text to appear (Harper and Row).

[47] L. LANZ, L. A. LUGIATO AND R. RAMELLA: Physica 54, 94 (1971); also M. A. ALBERTI, P. COHA-RAMASINA, G. RAMELLA: "On a Banach Space Formulation of Subdynamics in Quantum Statistics", preprint.

[48] P. J. M. BONGOARTS, M. FANNES, A. VERBEURE: "A Remark on Ergodicity, Dissipativity, Return to Equilibrium", preprint.

[49] M. DE HAAN AND F. HENIN: "Collision Operator in the Physical Representation, The Friedrichs Model", to appear in Physica.

[50] A. P. GRECOS: Physica 51, 50 (1971).

[51] R. BALESCU AND L. BRENIG: Physica 54, 504 (1971).

[52] P. A. M. DIRAC: Rev. Mod. Phys. 21 (1949).

[53] R. BALESCU AND R. KOTERA: Physica 33, 558 (1967).

[54] R. BALESCU: Physica 38, 119 (1968).

[55] A. W. SAENZ: Phys. Rev. 105, 546 (1957).

[56] H. MORI AND J. ROSS: Phys. Rev. 109, 1877 (1958).

[57] R. F. SNIDER: J. Chem. Phys. 32, 1051 (1960).

[58] D. K. HOFFMAN, J. MUELLER, C. F. CURTISS: J. Chem. Phys. 43, 2878 (1965).

[59] P. RÉSIBOIS: Physica 31, 645 (1965).

[60] G. SEVERNE: Trans. Th. and Stat. Phys. 1(2), 145 (1971).

[61] J. YVON: La Théorie Statistique des Fluides et l'Equation d'Etat (Herman, Paris, 1935).

[62] N. N. BOGOLIUBOV: "Problemi, Dynamitcheskij Theorie v Statistitcheskey Physikie", (Odis, Moskow, 1946); translated by E. K. Gora in J. de Boer and G. E. Uhlenbeck, ed. Studies in Statistical Mechanics, Volume I (North-Holland, Amsterdam, 1962).

[63] M. BORN AND H. S. GREEN: A General Kinetic Theory of Fluids (Cambridge University Press). See also H. S. GREEN: Molecular Theory of Fluids (North-Holland, Amsterdam, 1952).

[64] J. G. KIRKWOOD: J. Chem. Phys. 14, 180 (1946); 15, 72 (1947).

[65] L. WALDMAN: "On the Kinetic Equations with Internal Degrees of Freedom" in Acta Physica Austriaca Suppl. X The Boltzmann Equation ed. by E.G.D. Cohen and W. Thirring (Springer-Verlag, 1973), 223.

[66] J. J. N. BEENAKKER: "Non-Equilibrium Angular Momentum Polarization in Rotating Molecules", reference 65, p. 267.

[67] L. WALDMAN: "Transporterschienung in Gases von Mittleren Druch", in Handbuch der Physik ed. by S. Flugge 12 (Springer, Berlin, 1958).

[68] A. P. GRECOS AND W. C. SCHIEVE: Physica 46, 475 (1970).

[69] M. BARANGER: Phys. Rev. III, 481, 494 (1958); 112, 855 (1954).

[70] U. FANO: Phys. Rev. 131, 254 (1963).

[71] P. L. RONEY: "On the Theory of Combined Doppler and Binary Collision Foreign Gas Line Broadening", Thesis, Free University of Brussels, June 1973.

[72] S. CHAPMAN AND T. G. COWLING: The Mathematical Theory of Non-Uniform Gases (Cambridge University Press, New York, 1952).

[73] J. O. HIRSCHFELDER, C. F. CURTISS, R. B. BIRD: Molecular Theory of Gases and Liquids (John Wiley, New York, 1954).

[74] J. H. FERZIGER AND H. G. KAPER: Mathematical Theory of Transport Processes in Gases (North Holland, Amsterdam, 1972).

[75] A. TIP: Physica 52, 493 (1971).

[76] K. HAWKER AND W. C. SCHIEVE: Fourth Canadian Symposium on Theoretical Chemistry, Vancouver, July 1971.

[77] J. T. LOWRY: "Scattering Theory in Three-Body Quantum Kinetic Equations", Thesis, University of Texas, Austin, 1970; also J. T. LOWRY AND W. C. SCHIEVE: Trans. Th. and Stat. Phys. 1(3), 225 (1971).

[78] E.G.D. COHEN: in Fundamental Problems in Statistical Mechanics, ed. by C.G.D. Cohen (North-Holland, Amsterdam, 1962).

[79] M. C. GOLDBERGER AND K. M. WATSON: Collision Theory (John Wiley, New York, 1964).

[80] E. P. WIGNER: Phys. Rev. 40, 479 (1932).

[81] D. MASSIGNON: Mécanique Statistique des Fluides (Dunod, Paris, 1957).

[82] H. MORI, I. OPPENHEIM AND J. ROSS: in Studies in Statistical Mechanics I, J. de Boer and G. E. Uhlenbeck, ed. (North-Holland, Amsterdam, 1962).

[83] H. WEYL: The Theory of Groups and Quantum Mechanics, Second Edition (Dover Publ., New York, 1961).

[84] L. COHEN: J. Math. Phys. 7, 781 (1966).

[85] K. HAWKER: "Considerations of Formal Quantum Kinetic Theory", Thesis, University of Texas at Austin, June, 1974.

[86] D. BRASON: Phys. Rev. 135, 513; B 1255 (1964).

[87] S. IMAM-RAHAJOE AND C. F. CURTISS: J. Chem. Phys. 47, 5269 (1967).

[88] M. W. THOMAS AND R. F. SNIDER: J. Stat. Phys. 2, 2 (1970).

[89] E.C.G. STUECKELBERG: Helv. Phys. Acta 25, 577 (1952).

[90] J. WEINSTOCK: Phys. Rev. 132, 454 (1967); 140A, 460 (1965).

[91] J. R. DORFMAN AND E.G.D. COHEN: Phys. Letters 16, 124 (1965); J. Math. Phys. 8, 282 (1967).

[92] G. E. UHLENBECK: "The Validity and Limitations of the Boltzmann Equation", page 107, reference 65.

[93] E.G.D. COHEN: "The Generalizations of the Boltzmann Equation to Higher Order in the Densities", page 157, reference 65.

[94] T. NEAL: Phys. Rev. 169I, 508 (1968).

[95] P. RÉSIBOIS AND M. G. VELARDE: Physica 51, 541 (1971).

[96] J. V. SENGERS: "The Three Particle Collision Term in the Generalized Boltzmann Equation", page 177, reference 65.

[97] B. J. ALDER AND T. E. WAINWRIGHT: Phys. Rev. A1, 18 (1970).

TRANSFORMATION THEORY AND PHYSICAL PARTICLE DESCRIPTION OF DISSIPATIVE SYSTEMS

Claude George
Université Libre de Bruxelles
Belgique

CHAPTER I. INTRODUCTION

Much of the formal simplicity in the classical or quantum-mechanical description of reality comes from the existence of a theory of canonical transformations. The invariance under these transformations opens the possibility of choosing a representation such that the quantities involved appear in their simplest possible form.

It is, however, well-known that the validity of this transformation theory cannot easily be extended to systems with an infinite number of degrees of freedom in interaction. The major obstacle to this extension is the appearance of a new concept: dissipativity.

By formulating the problem in the superoperator language, Prigogine and the members of the Bruxelles group availed themselves from the beginning, of the formal similarity between the Liouville-von Neumann equation and the Schrödinger equation, in order to elaborate a theory of large dissipative systems [1].

As a first step, a kinetic description, valid in an asymptotic time limit, was proposed [2] which led afterwards to a "contracted description" in terms of "physical" entities [3]. More recently, this kinetic behaviour was understood as a projection of the complete evolution on a coherent subspace, containing all thermodynamics [4]. It appeared then that the evolutions in this subspace and in the complementary incoherent subspace were fully independent. (This procedure would correspond in ordinary classical mechanics to the separate consideration of invariants vs. other dynamical quantities, and in quantum mechanics to the singling out of the ground state vs. all excited states.)

The aim of the present paper is to insert most naturally these results in a more general framework, which will be seen to extend the transformation theory of quantum mechanics to systems studied in statistical mechanics.

It ought to be understood from what follows that this generalization was not looked for as such, but has transpired from already known results. We present it here because we think that it brings much formal clarity to the presentation, unites various concepts which have arisen at earlier stages and provides a useful guideline for further developments.

CHAPTER II. SCHRÖDINGER AND HEISENBERG REPRESENTATIONS

The formal solution of the Liouville-von Neumann equation

$$i\partial_t \rho(t) = L\rho(t) \tag{2.1}$$

can be written

$$\rho(t) = U(t)\rho \tag{2.2}$$

where ρ are the initial conditions, and U(t) the complete evolution operator.

Formally

$$U(t) = e^{-iLt} . \tag{2.3}$$

Then the average values at time t of a dynamical quantity A is

$$\langle A \rangle_t = T_r \, A^+ \rho(t) \equiv (A,\rho(t)) \tag{2.4}$$

and defines a "scalar product".

Together with this Schrödinger-like picture, a Heisenberg-like picture exists, which leads to the same physical results

$$\langle A \rangle_t = (A(t), \rho) \tag{2.5}$$

where

$$A(t) = U'(t) A . \tag{2.6}$$

U'(t) comes from the formal solution of the Heisenberg equation, which is known to differ from (2.1) by the mere replacement of L by L' = -L. U'(t) is then simply U(t) in which all L have been replaced by L'.

From the "scalar product" form of (2.4), the adjoint $U^+(t)$ to U(t) can be defined by partial integration

$$(A, U(t)\rho) = (U^+(t)A, \rho) \tag{2.7}$$

and the equivalence between Schrödinger and Heisenberg representations

(2.7) and (2.5) means that

$$U^{+}(t) = U'(t) \tag{2.8}$$

or

$$U(t) = U'^{+}(t) \equiv U^{*}(t) \tag{2.9}$$

We have introduced in (2.9) the *- operation, which combines the taking of the adjoint w. r. to the scalar product and a general change of signs of L. (2.9) then means that U(t) is *- hermitian.

Let us consider now a large system, with

$$L = L_{o} + \lambda \delta L \tag{2.10}$$

and for which our usual considerations apply [3].

The hermiticity property (2.9) valid for all times, remains then true in the kinetic limit [7]

$$\overset{(0)}{\Sigma}(t) = \overset{(0)}{\Sigma}{}^{*}(t) \tag{2.11}$$

and, in particular, the associated projector $\overset{(0)}{\Pi}$ is *- hermitian

$$\overset{(0)}{\pi} = \overset{(0)}{\pi}{}^{*} . \tag{2.12}$$

In the vectorial notation for the density matrix introduced in [3]

$$\rho = \begin{pmatrix} \rho_{0} \\ '\rho \end{pmatrix} \tag{2.13}$$

where ρ_{0} is the vacuum component and $'\rho$ the set of correlated components, the Σ and π operators take a block-matrix form

$$\overset{(0)}{\Sigma}(t) = \begin{pmatrix} e^{-i\Omega\psi t}A & e^{-i\Omega\psi t}AD \\ Ce^{-i\Omega\psi t}A & Ce^{-i\Omega\psi t}AD \end{pmatrix} \tag{2.14}$$

and

$$\overset{(0)}{\pi} = \begin{pmatrix} A & AD \\ CA & CAD \end{pmatrix} \tag{2.15}$$

(We refer to [3] for the details of the notations). A and $\Omega\psi$ are vacuum-diagonal but still operators in the occupation numbers; C and D are respectively asymptotic creation and destruction operators. As a consequence of (2.11) and (2.12) applied to (2.14) and (2.15), one obtains (see [7])

$$A = A^* \tag{2.16}$$

$$C = D^* \tag{2.17}$$

and

$$-\Omega\psi . A = A.(\Omega\psi)^* \tag{2.18}$$

or

$$(\overline{\Omega\psi}) = -(\Omega\psi)^* \tag{2.19}$$

where $\overline{\Omega\psi}$ has been defined in [8].

CHAPTER III. TRANSFORMATION THEORY

Let us introduce regular linear transformations on ρ and A

$$\rho = \Gamma\tilde{\rho} \tag{3.1}$$

such that the average values $\langle A\rangle$ remain unchanged

$$\langle A\rangle = (A,\Gamma\Gamma^{-1}\rho) = (\Gamma^{+}A,\Gamma^{-1}\rho) \equiv (\tilde{A},\tilde{\rho}) \tag{3.2}$$

The transformations introduced in [4] are of this type. $\tilde{\rho}$ consists there of the priviledged components

$$\tilde{\rho} = \begin{pmatrix} {}^{(0)}\rho_{0} \\ {}^{(\Lambda)}{}'\rho \end{pmatrix} \tag{3.3}$$

and the transformation operator is

$$\Gamma \rightarrow \Lambda = \begin{pmatrix} 1 & -D \\ C & \hat{1} \end{pmatrix} \tag{3.4}$$

The existence of the inverse operator

$$\Lambda^{-1} = \begin{pmatrix} A & AD \\ -CA & \hat{1}-CAD \end{pmatrix} \tag{3.5}$$

is a consequence of a relation between A, D and C operators [9], which can be written

$$1 + CD = A^{-1} \tag{3.6}$$

This transformation has been shown to diagonalize the complete evolution operator

$$U(t) = \Lambda \begin{pmatrix} e^{-i\Omega\psi t} & 0 \\ 0 & e^{-i\zeta t} \end{pmatrix} \Lambda^{-1} \tag{3.7}$$

which means that the priviledged components

$$\overset{(o)}{\rho}_0 = A\,[\rho_0 + \sum_\nu D_\nu\,\rho_\nu] \tag{3.8}$$

and

$$\overset{(\wedge)}{\rho}_\nu = \rho_\nu - C_\nu\,\overset{(0)}{\rho}_0 \tag{3.9}$$

evolve completely separately from each other, in what we have called the complementary $\overset{(0)}{\pi}$ and $\overset{(\wedge)}{\pi}$ subspaces.

Among all possible regular transformations (3.1 & 3.2) which can be considered, are those for which, ρ being related to $\tilde{\rho}$ through a functional of L, A is related to $\tilde{A}$ through the same functional, expressed in terms of L'.

$$\tilde{A} = \Gamma'^{-1}A = \Gamma^{+}A \tag{3.10}$$

These transformations are then *- unitary

$$\Gamma^{-1} = \Gamma^{*} \tag{3.11}$$

Clearly, the Λ- transformation (3.4) does not belong to that category, since

$$\Lambda^{*} = \Lambda'^{+} = \begin{pmatrix} 1 & C'^{+} \\ -D'^{+} & \hat{1} \end{pmatrix} = \begin{pmatrix} 1 & D \\ -C & \hat{1} \end{pmatrix} \neq \Lambda^{-1} \tag{3.12}$$

On the other hand, we have shown earlier [3] that a further transformation of the $\overset{(o)}{\rho}_0$ component was possible inside the $\overset{(0)}{\pi}$-space itself,

$$\overset{(R)}{\rho}_0 = \chi^{-1}\,\overset{(o)}{\rho}_0 \tag{3.13}$$

in order to obtain more symmetry in the kinetic description of the system. A similar transformation is possible inside the $\overset{(\wedge)}{\pi}$-space.

The idea we develop in the next paragraph is to avail ourselves of this additional freedom to bring the Λ-transformation into a *-unitary form by combining it with a regular transformation which does not mix the two subspaces $\overset{(0)}{\pi}$ and $\overset{(\wedge)}{\pi}$.

CHAPTER IV. STAR-UNITARY TRANSFORMATION

Let us introduce the "physical particle representation" through the transformation

$$\rho = \Lambda_p \overset{(p)}{\rho} \tag{4.1}$$

where Λ_p is a product of Λ (3.4) and a regular block-diagonal matrix

$$\Lambda_p = \Lambda \begin{pmatrix} \alpha^* & \circ \\ \circ & \beta^* \end{pmatrix} = \begin{pmatrix} \alpha^* & -D\beta^* \\ C\alpha^* & \beta^* \end{pmatrix} \tag{4.2}$$

When α and β are choosen such that

$$\alpha^* \alpha = A \tag{4.3}$$

and

$$\beta^* \beta = \hat{1} - CAD \tag{4.4}$$

it is verified that

$$\Lambda_p^{-1} \equiv \begin{pmatrix} \alpha^{*-1} & \circ \\ \circ & \beta^{*-1} \end{pmatrix} \Lambda^{-1} = \begin{pmatrix} \alpha & \alpha D \\ -\beta C & \beta \end{pmatrix} = \Lambda_p^* \tag{4.5}$$

and the transformation (4.1) is *- unitary.

Introducing the operators

$$\phi = \alpha^{*-1} \Omega\psi\alpha^* \tag{4.6}$$

and

$$\overset{(\hat{})}{\lambda} = \beta^{*-1} \xi\beta^* \tag{4.7}$$

it is obvious that the Λ_p-transformation also diagonalizes the complete evolution operator

$$U(t) = \Lambda_p \begin{pmatrix} e^{-i\phi t} & \circ \\ \circ & e^{-i\overset{(\hat{})}{\lambda} t} \end{pmatrix} \Lambda_p^{-1} \tag{4.8}$$

In addition, combining the *-unitarity of Λ_p (4.5) with the *- hermiticity of U(t) (2.9), one obtains

$$\phi^* = -\phi \qquad (4.9)$$

and

$$\overset{(\wedge)}{\lambda}{}^* = -\overset{(\wedge)}{\lambda} \quad . \qquad (4.10)$$

In the representation defined by (4.2), not only the evolution takes place in separate complementary subspaces, but also the density matrix and the observables are treated on equal footing.

Indeed, in $\overset{(0)}{\pi}$ e.g., the vacuum component of the density matrix

$$\overset{(p)}{\rho}_0 = \alpha\,[\rho_0 + \sum_\nu D_\nu \rho_\nu] \qquad (4.11)$$

is seen to evolve, in the Schrödinger picture, through the genuine collision operator ψ, while the vacuum components of the dynamical quantities are given by

$$\overset{(p)}{A}_0 = \alpha'[A_0 + \Sigma D'_\nu \, A_\nu] \qquad (4.12)$$

and the corresponding evolution operator in the Heisenberg picture is expressed in terms of ϕ'. Such a symmetry did not occur with the transformation Λ alone.

For a ρ entirely contained in the $\overset{(0)}{\pi}$-subspace (the canonical distribution is of this type: see [10]), one sees from (3.9) that

$$\rho_\nu = C_\nu \rho_0 \qquad (4.13)$$

and in the ϕ-representation, the only non vanishing component is

$$\overset{(p)}{\rho}_0 = \alpha[1 + D.C]\rho_0 = \alpha\, A^{-1}\rho_0 = \alpha^{*-1}\,\rho_0 . \qquad (4.14)$$

Similarly, for observables (like the total energy H) entirely contained in $\overset{(0)}{\pi}$, one has

$$\overset{(p)}{A}_0 = \alpha'[1 + D'.C']A_0 = \alpha^{+-1} A_0. \qquad (4.15)$$

In particular

$$\overset{(p)}{H}_0 = \alpha^{+-1} H_0. \qquad (4.16)$$

It has to be noticed that the factorization properties (4.3 and 4.4) do not lead to a unique *-unitary transformation: the diagonal blocmatrix in (4.2) is determined up to a *-unitary diagonal blocmatrix. In $\overset{(0)}{\pi}$ e.g.; if α is a solution of (4.3), $X\alpha$ is also a solution if

$$X^* = X^{-1}. \qquad (4.17)$$

A similar arbitrariness exists for β, and supplementary physical consideration will be needed in order to fix the Λ_p-transformation.

CHAPTER V. PROPERTIES OF THE Λ_p-TRANSFORMATION

Once the *-unitary transformation (4.2) is known, all the considerations, valid for ordinary unitary transformations can be formally transcribed in the present context.

Write

$$\Lambda_p^{-1} \equiv \begin{pmatrix} u \\ v \end{pmatrix} \tag{5.1}$$

with

$$u = (\alpha \quad \alpha D) \quad \text{and} \quad v = (-\beta C \quad \beta). \tag{5.2}$$

One can then associate with this transformation, the projectors u^*u and v^*v , which are nothing but $\overset{(0)}{\pi}$ and $\overset{(\hat{})}{\pi}$ respectively.

Indeed, e.g.,

$$u^*u = \begin{pmatrix} \alpha^* \\ C\alpha^* \end{pmatrix} (\alpha \quad \alpha D) = \begin{pmatrix} A & AD \\ CA & CAD \end{pmatrix} = \overset{(0)}{\pi} \tag{5.3}$$

The unitarity conditions for (5.1) give then immediately the decomposition and orthogonality properties of the projectors.

For instance, using (3.6) and (4.3), one can verify that

$$uu^* = \alpha(1 + DC)\alpha^* = 1 . \tag{5.4}$$

Similarly

$$vv^* = \beta(CD + \hat{1})\beta^* = \hat{1} \tag{5.5}$$

because of (3.6) and (4.4).

One finds also

$$uv^* = vu^* = 0 \tag{5.6}$$

and

$$u^*u + v^*v = I \tag{5.7}$$

The idempotence of $\overset{(0)}{\pi}$ and $\overset{(\wedge)}{\pi}$ is a straightforward consequence of (5.4) and (5.5) respectively, their orthogonality of (5.6) and their complementarity of (5.7).

Notice that the projector $\overset{(0)}{\pi}$ (hence $\overset{(\wedge)}{\pi}$) contain all of what we know about the complete *-unitary transformation Λ_p: it is expressed in terms of the operators A, D and C (which have been calculated in earlier work), whereas Λ_p contains α and β, determined up to a *-unitary factor only. In particular, for any choice of the *-unitary operator X in (4.18), $\overset{(0)}{\pi}$ (hence $\overset{(\wedge)}{\pi}$) remains the same.

We shall try to isolate a part of the Λ_p-transformation, which would be completely determined by the knowledge of $\overset{(0)}{\pi}$ only i.e. when A, D, C are given.

The exact Λ_p-transformation we are looking for is such that in particular

$$\overset{(p)}{H} = \Lambda_p'^{-1} H \tag{5.8}$$

possesses diagonal elements only (4.16).

If the hamiltonian was simply H_o, there would be no need for a transformation since in the original representation H_o is already diagonal. Λ_p is therefore a function of the coupling parameter λ which characterizes the strength of the interaction. Let us suppose that this λ-dependence is analytical.

Then one can always write formally the λ-derivative in terms of the operator itself

$$\partial_\lambda \Lambda_p^{-1} = \Lambda_p^{-1} M. \tag{5.9}$$

If $M(\lambda)$ is given, Λ_p^{-1} will be known, as for $\lambda = 0$, one has the boundary condition

$$\Lambda_p^{-1} (\lambda = 0) = I. \tag{5.10}$$

In order that Λ_p^{-1} be *-unitary, there is an evident condition on M

$$M^* = -M \ . \tag{5.11}$$

M remains unknown, but we may decompose it, following Mandel, into two parts

$$M = \xi + \Delta \tag{5.12}$$

the first one being given

$$\xi = \overset{(0)}{\pi}(\partial_\lambda \overset{(0)}{\pi}) - (\partial_\lambda \overset{(0)}{\pi})\overset{(0)}{\pi} \tag{5.13}$$

while the second one, Δ, is still unknown.

$\overset{(0)}{\pi}$ in (5.13), is considered, at the same time, as the projector corresponding to the exact Λ_p, solution of (5.9) and as the projector (2.15) expressed in terms of the operators A, D and C. An equivalent way of writing (5.13) is

$$\xi = \overset{(0)}{\pi}(\partial_\lambda \overset{(0)}{\pi}) + \overset{(\wedge)}{\pi}(\partial_\lambda \overset{(\wedge)}{\pi}) . \tag{5.14}$$

Because of the *-hermiticity of $\overset{(0)}{\pi}$ (2.12), one verifies from (5.13) that

$$\xi = - \xi^* \tag{5.15}$$

hence the condition (5.11) leads to the requirement on the unknown operator

$$\Delta^* = - \Delta \ . \tag{5.16}$$

In the two-component notation (5.1), (5.9) becomes then, using (5.12)

$$\begin{pmatrix} \partial_\lambda u \\ \partial_\lambda v \end{pmatrix} = \begin{pmatrix} u \\ v \end{pmatrix} (\xi + \Delta) = \begin{pmatrix} u\xi + u\Delta \\ v\xi + v\Delta \end{pmatrix} . \tag{5.17}$$

In view of the decomposition (5.3) of $\overset{(0)}{\pi}$, ξ (5.14) can be written

$$\xi = u^*u[(\partial_\lambda u^*)u + u^*(\partial_\lambda u)] - [(\partial_\lambda u^*)u + u^*(\partial_\lambda u)]u^*u \tag{5.18}$$

whence, using (5.4), and (5.6), one finds

$$u\xi = (\partial_\lambda u)\ (1 - u^*u) \tag{5.19}$$

$$v\xi = -v(\partial_\lambda u^*\ u) = (\partial_\lambda v)u^*u. \tag{5.20}$$

Then inserting (5.19), (5.20) into (5.17), one obtains

$$(\partial_\lambda u)\ \overset{(0)}{\pi} = u\Delta \tag{5.21}$$

and

$$(\partial_\lambda v)\ \overset{(\wedge)}{\pi} = v\Delta. \tag{5.22}$$

Multiplying (5.21) from the right by $\overset{(\wedge)}{\pi}$ and from the left by u^* one finds

$$0 = \overset{(0)}{\pi}\Delta\overset{(\wedge)}{\pi} \tag{5.23}$$

(5.23) means that the unknown operator Δ does not mix the components u and v in (5.17); it can therefore be written

$$\Delta = \overset{(0)}{\Delta} + \overset{(\wedge)}{\Delta} \tag{5.24}$$

with

$$\overset{(0)}{\Delta} = \overset{(0)}{\pi}\Delta\overset{(0)}{\pi} = -\overset{(0)}{\Delta}{}^* \tag{5.25}$$

and

$$\overset{(\wedge)}{\Delta} = \overset{(\wedge)}{\pi}\Delta\overset{(\wedge)}{\pi} = -\overset{(\wedge)}{\Delta}{}^*. \tag{5.26}$$

This further decomposition (5.24) leads then from (5.21) and (5.22) to the separate equations for u and v,

$$(\partial_\lambda u)\ \overset{(0)}{\pi} = u\ \overset{(0)}{\Delta} \tag{5.27}$$

$$(\partial_\lambda v)\overset{(\wedge)}{\pi} = v\overset{(\wedge)}{\Delta}. \tag{5.28}$$

One can similarly verify, from (5.14), using the idempotence of the projectors, that

$$\overset{(0)}{\pi}\,\xi\,\overset{(0)}{\pi} = \overset{(\wedge)}{\pi}\,\xi\,\overset{(\wedge)}{\pi} = 0. \tag{5.29}$$

The decomposition (5.12) of M is therefore done in terms of operators of two different types: one ξ which mixes the two subspaces $\overset{(0)}{\pi}$ and $\overset{(\wedge)}{\pi}$, and the other, still unknown, Δ which however operates inside each of the subspaces. It appears then as a natural decomposition, in view of our partial information on the exact transformation.

Due to the separation (5.27) and (5.28), we may in particular restrict ourselves to the examination of what happens in $\overset{(0)}{\pi}$ alone. In addition, we may factorize the solution u into a part u_1 [of the type (5.2) with a vacuum diagonal α_1] which will be independent of the choice of $\overset{(0)}{\Delta}$ and a *-unitary vacuum diagonal part X, which will directly depend on this choice

$$u = X.u_1. \tag{5.30}$$

Then (5.27) becomes

$$(\partial_\lambda X)u_1\overset{(0)}{\pi} + X(\partial_\lambda u_1)\overset{(0)}{\pi} = X\,u_1\overset{(0)}{\Delta}. \tag{5.31}$$

Let us choose u_1 to be the solution of

$$(\partial_\lambda u_1)\overset{(0)}{\pi} = 0 \tag{5.32}$$

with

$$u_1(\lambda = 0) = (1 \quad 0). \tag{5.33}$$

Then (5.32) can be written

$$[\ (\partial_\lambda \alpha_1) \qquad \alpha_\lambda(\alpha_1 D)\ ] \begin{pmatrix} A & AD \\ CA & CAD \end{pmatrix} = 0 \tag{5.34}$$

which leads to twice the same equation

$$(\partial_\lambda \alpha_1)A + \partial_\lambda(\alpha_1 D).CA = 0 \qquad (5.35)$$

or using (3.6)

$$\partial_\lambda \alpha_1 + \alpha_1(\partial_\lambda D).CA = 0. \qquad (5.36)$$

As X satisfies (4.18), $\overset{(0)}{\pi}$ will not depend on X,

$$\overset{(0)}{\pi} = u_1^* u_1 \qquad (5.37)$$

and (5.31) becomes, when multiplied to the right by u_1^*

$$\partial_\lambda X = X\, u_1 \overset{(0)}{\Delta} u_1^* \,. \qquad (5.38)$$

The property (5.25) is seen to guaranty the presupposed *-unitarity of X.

The equation (5.36) as it stands, can be easily shown to be equivalent to the equation proposed by Mandel [11] for the exact dressing operator in $\overset{(0)}{\pi}$. This corresponds then to X = 1, i.e., to the choice of a vanishing $\overset{(0)}{\Delta}$. Plausibility arguments in favor of this particular choice are presented elsewhere [12]. (See also the next paragraph).

A similar discussion for $\overset{(\wedge)}{\pi}$ will be presented in a series of forthcoming papers.

CHAPTER VI. COMPARISON WITH KNOWN RESULTS

In a non-dissipative system, using the unitary transformation in the Hilbert-space which diagonalizes the hamiltonian of the system

$$\tilde{H} = U^{-1} HU \tag{6.1}$$

one can transform the Liouville-von Neumann equation (2.1) where

$$L = [H, \quad] \tag{6.2}$$

into

$$i\partial_t \tilde{\rho} = \tilde{L} \tilde{\rho} \tag{6.3}$$

where

$$\tilde{L} = [\tilde{H}, \quad]. \tag{6.4}$$

In (6.3), the various components of

$$\tilde{\rho} = u^{-1} \rho U \tag{6.5}$$

evolve independently

$$i\partial_t \tilde{\rho}_0 = 0 \tag{6.6}$$

$$i\partial_t \tilde{\rho}_\nu = -\nu\tilde{\rho}_\nu \tag{6.7}$$

where ν is a notation for $(\tilde{H}_0^{N+\nu} - \tilde{H}_0^N)$.

To the unitary transformation U in the Hilbert space corresponds a unitary transformation $\mathbf{U} \otimes U^{-1}$ in the product space associated with the density matrix formalism, i. e., in the superoperator language. This transformation fulfills exactly the requirements [e.g., (3.2), (4.8), (4.16)] imposed on the Λ_p-transformation in the preceeding sections: it is therefore the Λ_p-transformation in the case of a non-dissipative system ($\Lambda_p^{-1} \equiv \gamma$ in [5]).

However, comparing (4.1) and (5.8) with (6.5) and (6.1) respectively, we see that there is no need here to distinguish between operators expressed as functionals of L and of -L.

For non-dissipative systems, then

$$\Lambda_p = \Lambda'_p \tag{6.8}$$

and the *-unitarity is no more distinct from the ordinary unitarity.

In particular

$$A = A' \tag{6.9}$$

$$\alpha = \alpha' \tag{6.10}$$

and

$$C = D^+ . \tag{6.11}$$

The transformation $U \otimes U^{-1}$ has an additional property: it preserves the algebra of the operators in the Hilbert space. If A and B are two observables, one verifies from (6.1) that

$$\overset{(p)}{A} \overset{(p)}{B} = \Lambda_p^{-1} (AB). \tag{6.12}$$

The invariants lie all in $\overset{(0)}{\pi}$ [10], and can be diagonalized simultaneously with H, so that, for these operators, one can restrict (6.12) to the diagonal components in the p-representation:

$$\overset{(p)}{A}_0 \overset{(p)}{B}_0 = \alpha^{+-1} (AB)_0 . \tag{6.13}$$

In particular, for any power of H itself, one has

$$(\overset{(p)}{H}_0)^n = \alpha^{+-1} (H^n)_0 . \tag{6.14}$$

In going to the dissipative systems, one does not expect (6.12) to remain valid for each couple of observables. However (6.14) is precisely the kind of relation we would need in order to obtain, at equilibrium, a purely combinatorial form of the entropy (in the physical particle representation). On these physical grounds, we have proposed in [3] to impose (6.14) to remain true for dissipative systems, in order to raise the unitary uncertainty of the dressing operator α. (See also Mandel [5] and Turner [6].) It has, however been verified since that this condition was not fulfilled, and the dissipative extension of (6.14) remains an open question.

Explicit calculations on a formal dissipative model [13] have shown that, when a certain condition was satisfied, the only places where the change $L \to L' = -L$ could affect the transformation were the asymptotic creation and destruction operator C,D. This reality condition on diagonal matrix elements of powers of the perturbation is verified in all the physically interesting cases we have considered.

With this condition fulfilled, (6.9) and (6.10) remain valid, even for dissipative systems, but (6.11) does not and is replaced by (2.17). The collision operator is also invariant under this change

$$\Omega\psi = \Omega'\psi' \quad . \tag{6.15}$$

We see therefore, that in $\overset{(0)}{\pi}$, the dissipative Λ_p comes still closer to the usual unitary transformation, as it is in C and D alone that we have to distinguish between expressions in terms of L and -L.

The distinction disappears in products of D and C giving diagonal operators [see (3.6)], even when one of them is differentiated with respect to λ, as in (5.36) [5].

In view of these remarks, using (6.9) and (6.10), and setting

$$\alpha = X^{+} \tag{6.16}$$

one recovers from (4.3) the original decomposition introduced in [3]

$$A = XX^{+} \; . \tag{6.17}$$

Then, (4.17) becomes

$$\overset{(p)}{H}_{0} = X^{-1} H_{0} \equiv H_R \tag{6.18}$$

(H_R is the notation of [3]); (4.6) and (4.9) give

$$\phi = X^{-1}\Omega\psi X = -\phi^{+} \tag{6.19}$$

and the whole scheme presented as the "contracted description" is recovered in $\overset{(0)}{\pi}$.

The Mandel equation for the dressing operator X^{-1} [11] is then simply derived from (5.36):

$$\partial_\lambda X^{-1} = X^{-1} \, \mathrm{AD}.(\partial_\lambda C) \; . \tag{6.20}$$

CHAPTER VII. GENERAL CONCLUSIONS

The evolution of a large dissipative system is most conveniently described in the physical representation we have introduced in Chapter V.

We have seen in Chapter VI that the transition to this representation is the natural generalization of the unitary transformation which, in the non-dissipative system, diagonalizes the total Hamiltonian and gives to the evolution its simplest possible form.

Indeed, the scalar product (2.4) is known to be invariant under the usual unitary transformations of ordinary quantum mechanics

$$(\Gamma A,\ \Gamma\rho) = (A,\ \rho) \qquad (7.1)$$

with

$$\Gamma^{+}\Gamma = 1 \qquad (7.2)$$

in which Γ acts in the same (factorizable) way on both states and observables

$$\Gamma\rho = U\rho U^{+} \qquad \text{and} \qquad \Gamma A = UAU^{+} \qquad (7.3)$$

but becomes singular for dissipative systems.

Now, the scalar product (2.4) is also invariant under a larger class of transformations, which, through different realizations, operate differently on states and observables, but still preserve the average values and the normalization

$$\Gamma[\langle A\rangle] = (\Gamma(-)A, \Gamma(+)\rho) = (A, \rho) \qquad (7.4)$$

with

$$\Gamma^{+}(-)\ \Gamma(+) = 1 \qquad (7.5)$$

and

$$\Gamma(\pm)\ 1 = 1. \tag{7.6}$$

The transformation Λ_p, which leads from a "bare particle representation" to a "physical particle representation" is precisely of this type: when acting on states or observables, it differs by the sign of L.

The *-operation consists then in the composition of the usual hermitian conjugation with respect to the scalar product (2.4) and of the transition from one realization to the other

$$\Gamma^{+}(-) = \Gamma^{*}(+) \ . \tag{7.7}$$

In the presence of dissipation, a consequence of the choice (4.2) for the *-unitary transformation Λ_p, is that the thermodynamical behaviour (in $\overset{(0)}{\pi}$) is automatically separated from the incoherent one (in $\overset{(\wedge)}{\pi}$), which will not contribute appreciably to average values of quantities of macroscopic interest.

Moreover, the kinetic evolution in $\overset{(0)}{\pi}$ is now expressed in terms of an operator ϕ of a generalized Boltzmann type, and is thus due to <u>collisions</u> (a purely dissipative concept) <u>between physical entities</u> (in terms of which the Hamiltonian would be diagonal in the absence of dissipativity). In this sense, a Hamiltonian formulation and dissipativity are reconciled. Our choice of the determination for the dressing in $\overset{(0)}{\pi}$ can be characterized by the constant care of deviating as little as possible from the non-dissipative case. A similar choice has now been made for $\overset{(\wedge)}{\pi}$, on similar grounds. We refer to forthcoming publications for a study of the relations which exist between the physical entities defined in $\overset{(0)}{\pi}$ and the evolution in $\overset{(\wedge)}{\pi}$.

ACKNOWLEDGEMENTS

The author is grateful to Professor I. Prigogine for his stimulating interest.

This work was first presented in a series of seminars at the University of Texas at Austin, during fall 1969. It is a pleasure to thank all the members of the Center for Statistical Mechanics and Thermodynamics for their hospitality. Fruitful discussions with Professor Henin, Dr. Mandel and J. W. Turner are also acknowledged.

References:

[1] I. PRIGOGINE: Non Equilibrium Statistical Mechanics, (New York: Interscience, 1962).

[2] P. RESIBOIS: in Physics of Many Body Systems (ed. Meeron), (Gordon and Breach, 1966).

[3] C. GEORGE: Physica 37, 182 (1967).

[4] I. PRIGOGINE, C. GEORGE, F. HENIN: Physica 45, 418 (1969).

[5] P. MANDEL: Physica 48, 397 (1970).

[6] J. W. TURNER: Physica 51, 351 (1971).

[7] P. MANDEL: Bull. Acad. Belg. Cl. Sc. 54, 949 (1968).

[8] C. GEORGE: Bull. Acad. Belg. Cl. Sc. 53, 623 (1967).

[9] C. GEORGE: Bull. Acad. Belg. Cl. Sc. 56, 386 (1970).

[10] R. BALESCU, P. CLAVIN, P. MANDEL, J. W. TURNER: Bull. Acad. Belg. Cl. Sc. 55, 1055 (1969).

[11] P. MANDEL: Physica 50, 77 (1970).

[12] I. PRIGOGINE, C. GEORGE, F. HENIN, P. MANDEL, J. W. TURNER: Proc. Nat. Acad. Sci. (U.S.A.) 66, 709 (1970).

[13] C. GEORGE: Physica 39, 251 (1968).

KINETIC THEORY OF GASES IN GENERAL RELATIVITY THEORY

Jürgen Ehlers
Max-Planck-Institut für Physik und Astrophysik
München

CHAPTER I. INTRODUCTION

The purpose of this lecture is, firstly, to describe the framework of the general-relativistic kinetic theory of gases and, secondly, to sketch some of the advances which have been made in this field during the last few years. Systematic expositions containing details and proofs can be found in references [1]- [6] and [29].

Some of the reasons for developing a general-relativistic kinetic theory of gases are the following. The traditional fluid description for the sources of gravitational fields does not seem to be appropriate in some cases of astrophysical interest such as stellar systems or the "galaxy-gas" of cosmology, since collisions are rare and the mean free paths are long. Also, a fluid description does not provide values for transport and reaction coefficients, whereas the less phenomenological kinetic theory does. Moreover, radiation (photons, neutrinos) can be described as a gas of zero-mass particles for some purposes, and only a relativistic version of kinetic theory can provide a unified treatment of such gases and ordinary gases. Also, relativistic kinetic theory helps clarifying controversial questions of relativistic thermodynamics. Finally, the relativistic version of kinetic theory is in some respects simpler and more transparent than its nonrelativistic predecessor; here as in other branches of Physics the unifying and simplifying power of the spacetime - geometrical point of view first put forward by H. Minkowski is clearly visible.

CHAPTER II. REMARKS ABOUT GENERAL RELATIVITY THEORY

In Einstein's theory of gravitation spacetime, the arena of all physical processes, is assumed to be a four dimensional manifold which carries a pseudoriemannian metric. The metric tensor g_{ab} can locally be transformed to the Minkowski-form g_{ab} = diag. (1,1,1,-1). It determines the light-cones, the distinction between time-like, space-like and null (or light-like) vectors, it defines the causal structure of spacetime, and it establishes (part of) the connection between the mathematical formalism and Physics by providing definitions of (proper) times and distances. At the same time, the ten functions $g_{ab}(x^c)$ which, in the presence of inhomogeneous gravitational fields, cannot be transformed into constants by coordinate transformations in finite regions, act as potentials of the gravitational field. In a weak, quasistationary field, e.g., one has approximately

$$ds^2 = g_{ab}dx^a dx^b \approx d\vec{x}^2 - (1 + 2\frac{U}{c^2})c^2dt^2, \qquad (0)$$

where U is the Newtonian gravitational potential, and in more general situations all ten g_{ab}'s contribute to the field.

Just as in Newtonian theory the potential U is related to the mass density ρ of matter by Poisson's equation $\nabla^2 U = 4\pi G\rho$, so in Einstein's theory of gravitation the metric field g_{ab} is coupled to matter by the field equation

$$G^{ab} = T^{ab}. \qquad (1)$$

Here, the Einstein tensor G^{ab} is a symmetric second-rank tensor constructed from the g_{ab}'s and their first and second derivatives, and T^{ab} is the stress-energy-momentum tensor of all the matter (particles and non-gravitational fields) present. Here and in the sequel, the convention $G = \frac{1}{8\pi}$, c = 1 is used; later we shall also put k (Boltzman's constant) = 1.

Equation (1) implies the energy-momentum balance equation

$$T^{ab}{}_{;b} = 0 \qquad (2)$$

in which $(\ )_{;b}$ denotes covariant differentiation with respect to x^b. In a gravitational field, eq. (2) is no longer a local conservation law, but expresses the response of matter to gravity; it restricts (and in simple cases determines) the motion of bodies. As will be indicated later, eq. (2) can be derived from simpler assumptions in kinetic theory, independently of the field equation (1).

To solve Einstein's field equation (1) means, apart from specifying a manifold which serves as the domain for the tensors g_{ab}, T^{ab} etc., to choose a physically reasonable model of matter which specifies the form of T^{ab} in terms of matter or field variables (and of g_{ab}), and then to find values for the metric field and the matter variables which satisfy the ten coupled, quasilinear (but nonlinear!) differential equations (1), possibly in conjunction with further, non-gravitational, laws describing the sources. In general, neither the left-hand side nor the right-hand side of eq. (1) can be considered as given; one is faced with the problem of finding "simultaneously" all the quantities g_{ab}, T^{ab} etc. such that they satisfy eq. (1) "selfconsistently".

In macroscopic applications of general relativity theory the standard model of matter has been the perfect fluid, given by its energy density μ, its (isotropic) pressure p, and its 4-velocity u^a (a timelike unit vector tangent to the streamlines); for it

$$T^{ab} = \mu u^a u^b + p(g^{ab} + u^a u^b). \qquad (3)$$

In this case, eq. (2) is equivalent to a continuity equation for μ and a generalized Euler equation for u^a. Just as in Newtonian theory the system of equations (1), (3) is underdetermined; the simplest way to obtain a system such that Cauchy initial data uniquely determine the future evolution is to add a relation

$$p = f(\mu) \tag{4}$$

between pressure and density.

This model of matter is a very special one. Thermal phenomena are neglected in (3) and (4); in particular, no transport phenomena are taken into account. Although these drawbacks can be removed partly at the phenomenological level, the choice of non-equilibrium equations remains a matter of guesswork, and no transport coefficients are given. Moreover, if the matter of interest is radiation, a description like that in eq. (3), even if more or less correct, does not give sufficiently detailed information, since one would like to bring into the picture the spectrum of the radiation.

One simple way to improve the description of matter is to turn to kinetic theory, as will be done now.

CHAPTER III. BASIC CONCEPTS AND LAWS OF RELATIVISTIC KINETIC THEORY

The theory to be outlined in this section was developed in small pieces over a long period of time. The main steps have been taken by Jüttner (1911, 1928), Synge (1934), Walker (1936), Lichnerowicz and Marrot (1940), Chernikov (1960 - 1963), Tauber and Weinberg (1961), and Ehlers (1961).* Papers concerned with applications, approximation methods, special solutions etc. will be mentioned in section IV; no attempt is made, however, to give a complete list of references.

The assumptions on which the kinetic theory of gases is based are the following:

(a) The interactions between the particles constituting the gas can be divided into <u>long range forces</u> and <u>weak</u>, <u>short range forces</u> such that

$(a)_1$ the long range forces can be accounted for in terms of a <u>mean field</u> generated collectively by the particles of the gas through <u>macroscopic field equations</u>, and

$(a)_2$ the short range forces can be taken into account in terms of (elastic or inelastic) <u>point-collisions</u> whose probability of occurence is governed by cross-sections taken from a special-relativistic scattering theory.

In accordance with this, it is assumed that

(b) between collisions, particles move like <u>test particles</u> in the mean field.

Finally, the usual assumption is made that

(c) the pattern of world-lines and collision events may be treated as a <u>random structure</u> whose (physically relevant) properties can be described by <u>smooth expectation values</u>.

---

* See references [7],[8],[9],[10],[1], [11], [12], respectively.

These assumptions are physically plausible for dilute gases. Their justification from first principles of many-particle dynamics is a formidable problem which is not attempted here; rather, we follow Boltzmann in formulating directly laws in a suitably defined one-particle phase space which seem reasonable under the above assumptions.

As the only long range interaction we shall here take gravitation; electromagnetic fields can easily be included in an analogous way. As short range interactions we have in mind non-gravitational interactions such as electromagnetic multipole forces or nuclear forces.

Let the gas consist of particles of proper mass m (≥ 0). Between collisions, we have according to (b) geodesic motion, i.e.

$$\frac{dx^a}{dv} = p^a, \quad \frac{Dp^a}{dv} = \frac{dp^a}{dv} + \Gamma^a_{bc}\, p^b p^c = 0. \tag{5}$$

The parameter v is chosen such that p^a is the 4-momentum, thus

$$p_a p^a = -m^2. \tag{6}$$

$\frac{D}{dv}$ indicates the absolute derivative; the quantities

$$\Gamma^a_{bc} = \frac{1}{2} g^{ad} (g_{db,c} + g_{dc,b} - g_{bc,d}) \tag{7}$$

form the components of the Riemannian connection associated with g_{ab}. Physically, these quantities are the relativistic analogues of the components of the gravitational field strength. Their non-tensorial character is (physically) due to the principle of equivalence.+ The g_{ab} and Γ^a_{bc} represent the mean field according to assumption $(a)_1$.

+ The Γ^a_{bc}'s form a good example of an object which is neither a tensor nor a spinor, but nevertheless of fundamental geometrical and physical importance.

In nonrelativistic kinetic theory it is customary to represent the states of particles as points in a six-dimensional $(\vec{q}, \vec{p})$ phase space; these points move in the course of time. Such a description refers to a particular inertial frame of reference. If one passes from one inertial frame to another one, the phase-space description (of a particular gas state) changes in a simple and obvious manner. A similar description is still possible in special relativity, and has in fact been employed (e.g., by Jüttner). Here, the change connected with a change of the inertial frame is already more complicated due to the relativity of simultaneity. In general relativity, inertial frames (in finite domains) do not exist due to the very nature of gravitational fields*. Hence, the above description cannot be taken over without essential changes.

The best plan to overcome this difficulty is, not to use arbitrary, non-inertial frames of reference with a necessarily highly arbitrary splitting of spacetime into "space" and "time", but rather to look for the frame-independent meaning of the ordinary phase-space description, which can then be carried over to general relativity almost without change.

In geometric language the phase space description amounts to the following: In spacetime $X = \{\vec{x},t\}$, the motion of a particle is represented as a worldline $(\vec{x}[t], t)$. The instantaneous state of a particle with mass m can be specified by an event $(x[t_0], t_0)$ and a world-momentum $(m\dot{\vec{x}}[t_0], m)$ at that event. The collection of all possible instantaneous states (for fixed m) is a seven-dimensional manifold M, the augmented phase space. A mean (or external) field defines in M a family of curves, the phase flow; it represents all

* If inertial frames are to be identifiable locally by means of mechanical experiments or if unbounded matter distributions are considered (as in cosmology), this statement holds already in Newtonian theory.

possible test particle orbits, "lifted" from X to M. The six-dimensional manifold $\dot{M}$ of phase orbits (obtained from M by identifying points contained in the same orbit) is the intrinsic object corresponding to the many $(\vec{q}, \vec{p})$ phase spaces associated with the various inertial frames.

One can assign a <u>size</u> to a tube T of phase orbits by intersecting T with a hypersurface t = const. and forming the Lebesgue-measure $\int d^3x\, d^3p$ of that intersection, using inertial coordinates. According to <u>Liouville's theorem</u>, this size is independent both of the inertial frame used to compute it and of the instant t defining the cross section. Thus, the Lebesgue measure defines a <u>measure ω on $\dot{M}$</u>.

The description just given applies to (special and) general relativity immediately. The augmented phase space is here given by

$$M = \bigcup_{x \in X} P_x, \tag{8}$$

the union of all the <u>mass-shells</u> $P_x = \{p^a\colon\ p_a p^a = -m^2\}$ belonging to the events* x of X. (Coordinates in M are x^a, p^ν, where a = 1,...., 4 and ν = 1, 2, 3. p^4 is fixed by the mass shell condition (6).) The phase flow in M is determined by equation (5). A measure on $\dot{M}$ is obtained as follows. First, form the product $\Omega = \eta \wedge \pi$ of the Riemannian measure $\eta = \sqrt{-\det(g_{ab})}\, d^4x$ of X with the measure $\pi = \frac{d^3p}{p^4}$ (in orthonormal coordinates at x) of P_x, obtaining a measure on M which can be shown to be invariant under the phase flow (5). Then, contract† Ω with the vector field+

---

* P_x cannot be identified with P_y for $x \neq y$ since parallel transport in X is not integrable. M is a <u>fibre bundle</u> over X, but not a cartesian product.

† By definition $(L\cdot\Omega)_{b_1\ldots b_6} = L^a\Omega_{ab_1\ldots b_6}$; L^a, $\Omega_{a_1\ldots a_7}$ being the components of L and Ω, respectively.

\+ We follow the usage of modern differential geometry of identifying a vector with its directional derivative operator.

$$L = p^a \frac{\partial}{\partial x^a} - \Gamma^{\nu}_{bc} p^b p^c \frac{\partial}{\partial p^\nu} \qquad (9)$$

(on M) to obtain a six-form $\omega = L \cdot \Omega$ which, again, can be shown to be invariant under the phase flow generated by the vector field (9). The measure of any region in $\dot{M}$, i.e., any tube T of phase orbits in M, is then defined (in strict analogy to the Newtonian case discussed above) as $\int \omega$, taken over any cross section of T.

The state or, rather, the history of an individual gas can be described by specifying those segments of phase orbits which are occupied by particles. It follows* from assumption (c) that the (average) number of occupied states (≡ phase orbit segments) intersecting a hypersurface H of M can be expressed as an integral

$$N[H] = \int_H f\omega, \qquad (10)$$

where $f = f\,(x,p)$ is a non-negative, smooth, scalar function on M, called the (one particle) distribution function. Since ω coincides for a local observer with the ordinary phase-element $d^3 x\, d^3 p$, f has, for each such observer, the same meaning as the distribution function in nonrelativistic kinetic theory.

A collision $(x;\, p,\bar{p} \to p',\, \bar{p}')$ at x gives rise, in M, to two endpoints (x,p), $(x,\bar{p})$ of occupied phase orbit segments to be called annihilations, and two initial points, (x,p'), $(x,\bar{p}')$, called creations. Counting the latter ones positively, the former ones negatively, one can easily deduce from (10) that the density of collisions in M with respect to the measure Ω is given by

$$L(f) = p^a \frac{\partial f}{\partial x^a} - \Gamma^{\nu}_{bc} p^b p^c \frac{\partial f}{\partial p^\nu}\,. \qquad (11)$$

Hence, the spacetime density of collisions at x in which particles with 4-momenta in π are created, is given by $L(f)\pi$; this is the

---

* For a rigorous formulation and a proof, see [5a], [5b].

ordinary collision rate.*

The preceding remark implies: Absence of collisions or, more generally, detailed balance between creations and annihilations (direct and inverse collisions) is expressed by

$$L(f) = 0, \tag{12}$$

the "collisionless" Boltzmann equation. (In the case of weak, quasistationary gravitational fields and slowly moving particles eqs. (0), (7), (11), and (12) reduce to the well known gravitational Vlasov equation

$$\frac{\partial f}{\partial t} + \frac{\vec{p}}{m} \cdot \frac{\partial f}{\partial x} - m\nabla U \cdot \frac{\partial f}{\partial \vec{p}} = 0.)$$

It is apparent from the meaning of f that the moments

$$N^a(x) = \int_{P_x} p^a f\pi, \tag{13}$$

$$T^{ab}(x) = \int_{P_x} p^a p^b f\pi, \tag{14}$$

represent currents in spacetime. N^a is the particle 4-current density (also called numerical flux), and T^{ab} is the kinetic stress energy momentum tensor of the gas described by f. In a similar way higher order moments can be defined.

Since N^a is timelike, one can factor it into a non-negative scalar n and a timelike unit vector u^a;

$$N^a = n\,u^a. \quad (u_a u^a = -1) \tag{15}$$

An observer travelling with 4-velocity u^a will observe no particle flux and will measure the particle density n. Hence, in accordance with nonrelativistic terminology one might call u^a the mean 4-velocity of the gas, and n, the proper particle density.

---

* See, e.g., references [2], [4], [5].

Any tensor T^{ab} constructed via eq. (14) can be decomposed uniquely as*

$$T^{ab} = \bar{\mu}\, \bar{u}^a\, \bar{u}^b + \bar{p}^{ab} \qquad (16)$$

with

$$\bar{\mu} \geq 0, \quad \bar{u}_a \bar{u}^a = -1, \quad \bar{p}_{ab} \bar{u}^b = 0 \ . \qquad (17)$$

An observer travelling with 4-velocity $\bar{u}^a$ would, consequently, find the gas to have a vanishing momentum density. Thus, $\bar{u}^a$ could also be considered to be "the" mean 4-velocity of the gas. In general, however, $\bar{u}^a \neq u^a$. The physical reason for this is, of course, the velocity-dependence of the (relative) inertial mass of a particle. For clarity, u^a is called the <u>kinematical mean 4-velocity</u>, and $\bar{u}^a$ is called the <u>dynamical mean 4-velocity</u> of the gas (Synge 1956, [13]). (For a multicomponent gas the ambiguity in the choice of a mean 4-velocity is even greater; to avoid confusion, it is necessary to define precisely which mean velocity is used in a particular context.)

With respect to <u>any</u> mean 4-velocity u^a, T^{ab} can be uniquely decomposed according to†

$$T^{ab} = \mu u^a u^b + p(g^{ab} + u^a u^b) + 2u^{(a} q^{b)} + \pi^{ab} \qquad (18)$$

with

$$\mu \geq 0, \ u_a u^a = -1, \ u_a q^a = 0, \ \pi_{ab} u^b = 0, \ \pi^a{}_a = 0. \qquad (19)$$

The quantities μ, p, q^a, π^{ab} represent the <u>energy density</u>, <u>mean kinetic pressure</u>, <u>energy current density</u>, and <u>shear viscosity</u> with respect to u^a, respectively. We shall henceforth choose the u^a in (18) to be the kinematical mean 4-velocity; then q^a is also called

---

* Synge, [13], see also [5a], [5b].

† By definition, $u^{(a} q^{b)} = \frac{1}{2} (u^a q^b + u^b q^a)$.

the heat flux. $q^a = 0$ if and only if $\overline{u}^a = u^a$; this property serves to define adiabatic processes.

Consider a spatially bounded gas, such as a gas enclosed in a container or (in good approximation) a star. An "instant of time" is represented in relativity as a spacelike hypersurface G in X. G defines a hypersurface $\hat{G} = \{(x,p): x\in G,\ p\in P_x\}$ in the augmented phase space M. The entropy of the gas at the instant G is defined to be

$$S[G,f] = -\int_{\hat{G}} f \log f\ \omega. \tag{20}$$

This expression is the straightforward generalisation of the corresponding one in the nonrelativistic theory. It can be motivated either by adapting Boltzmann's counting procedure to the relativistic setting*; or by using a quantum model of a gas, starting from the definition S = - trace (W log W) of its entropy in terms of its statistical operator W, and re-expressing that by means of correspondence arguments in terms of classical quantities†. (The second procedure gives, of course, also the expressions appropriate to Bose or Fermi gases, both of which reduce essentially to (20) in the nondegenerate limit.)

If the form of the measure ω (which has been described above) is taken into account, it follows that (20) can be rewritten in the form

$$S[G,f] = \int_G S^a \sigma_a, \tag{21}$$

where

$$S^a(x) = -\int_{P_x} p^a f \log f \pi \tag{22}$$

is a vector field in X and σ_a is the standard hypersurface element of G. S^a is called the entropy flux of the gas.

---

* See, e.g., references [7a], [14].

† See reference [5a].

In order to obtain a time evolution equation for f one can carry over to relativity Boltzmann's collision hypothesis. Considering again a simple gas with elastic binary collisions only - other cases may be treated similarly - and remembering assumption $(a)_2$ and the meaning of L(f), one gets the <u>Boltzmann equation</u> ([1], [10], [11], [12])

$$L(f) = \int(\bar{f}\,\bar{f}' - f\,f')\, W\delta(\Delta p)\pi'\hat{}\bar{\pi}\hat{}\bar{\pi}', \tag{23}$$

where the usual abbreviations $f' = f(x,p')$ etc. have been employed and the nonnegative Lorentz invariant function $W(p,p',\bar{p},\bar{p}')$ is related to the differential scattering cross section $\sigma(E,\theta)$ by (see refs. [3],[5],[32IV])

$$W\delta(\Delta p)\bar{\pi}\hat{}\bar{\pi}' = E\left(\left(\frac{E}{2}\right)^2 - m^2\right)^{\frac{1}{2}} \sigma(E,\theta)\, d\bar{\Omega}. \tag{24}$$

($E = [-(p + p')^2]^{\frac{1}{2}}$ is the total CM-energy, θ the CM scattering angle, and $d\bar{\Omega}$ is a solid angle which refers to the direction of $\bar{p}$ in the CM frame of collision $[p,p' \to \bar{p},\bar{p}']$.) The factor $\delta(\Delta p) = \delta(\bar{p} + \bar{p}' - p - p')$ accounts for conservation of 4-momentum during collisions.

Whereas the left-hand side of eq. (23) is essentially general-relativistic (see eq. (11)), the right-hand side is essentially special-relativistic; this conforms to the assumptions $(a)_1$ and $(a)_2$ stated above.

The Boltzmann equation implies the <u>particle conservation law</u>

$$N^a{}_{;a} = 0 \tag{25}$$

and the <u>4-momentum balance equation</u> (2), as is seen by differentiating covariantly eqs. (13) and (14) and using eq. (23). Similarly, differentiation of the entropy flux (eq. (22)) and use of eq. (23) leads to

$$S^a{}_{;a} \geq 0, \qquad (26)$$

the relativistic version of Clausius's inequality (H-theorem) [11], [12]. The quantity on the left-hand side of (26) is the (invariant) entropy production rate.

The last inequality can be used to motivate the definition of local equilibrium distributions as those distributions which, at an event x, have a vanishing entropy production rate. Using again the Boltzmann equation one can show*, as in nonrelativistic theory, that f has the stated property if and only if log f is an additive collision invariant. This in turn implies that f must have the form*

$$f = e^{\alpha + \frac{u_a p^a}{T}}, \qquad (27)$$

where $T > 0$, and u^a is a timelike unit vector†. Eq. (27) gives the relativistic analog of the Maxwell-Boltzmann distribution, first derived by probabilistic arguments by Jüttner (ref. [7a]). If (27) is combined with egs. (13), (14), and (22), then equations (15), (3) and $S^a = su^a$ follow, with well-determined functions $n(\alpha,T)$, $\mu(\alpha,T)$, $p(\alpha,T)$, $s(\alpha,T)$, together with standard thermodynamic relations which identify T as the (absolute) temperature, and $\alpha = \frac{\kappa}{T}$ + const. where κ is the chemical potential.

Global equilibrium in a spacetime domain D requires f to have the form (27), with functions $\alpha(x)$, $T(x)$, $u^a(x)$ such that the Boltzmann equation holds. It turns out that this is true precisely if α = const. and

---

* For the most careful treatment of these points, see ref. [2a].

† If the right hand side of eq. (23) is modified so as to account for the Pauli principle or stimulated scattering, the analogous reasoning leads to the relativistic Fermi-Dirac and Bose-Einstein distributions, respectively ([11]; see also [4],[5]).

$$\left(\frac{u_a}{T}\right)_{;b} + \left(\frac{u_b}{T}\right)_{;a} \quad \propto \quad \begin{cases} 0 \text{ if } m > 0 \\ g_{ab} \text{ if } m = 0 \end{cases} \tag{28}$$

This means that $\frac{u^a}{T} = \xi^a$ must generate a group of congruent (if $m > 0$) or conformal (if $m = 0$) mappings $x^a \to x^a + \varepsilon\,\xi^a$ of spacetime into itself, and that T (and also κ) must vary in D just like $(-\xi_a \xi^a)^{-1/2}$. That is, for particles with positive rest mass m global equilibrium is possible only in a stationary spacetime*, and then in stationary coordinates the temperature varies according to Tolman's law

$$T\,\sqrt{-g_{44}} = \text{const.} \tag{29}$$

(x^4 = time coordinate, $g_{ab,4} = 0$). This means that the temperature depends on the gravitational potential in such a way that the gravitational redshift of photons does not disturb the equilibrium set up by exchange of radiation. ($-g_{44} \simeq c^2 + 2U$, see eq. (0).)

If $m = 0$, equilibrium is compatible with certain nonstationary states of the gravitational field. An important example is provided by black body radiation in an isotropically expanding space; this is the current model for the well-known 3°K cosmic fireball radiation. In this case, eq. (28) says that the radiation temperature drops like the inverse of the "world radius".

The fundamental equations for a gravitating gas (according to kinetic theory) are the Einstein field equation (1) with a source term as given by eq. (14), coupled with the Boltzmann equation (23). (Generalisations to gas mixtures, or to Fermion or Boson gases require obvious modifications.) Since both equations seperately imply eq. (2), it appears that they are compatible, and that the

---

* In nonrelativistic kinetic theory, distributions without entropy production are possible even in some non-stationary fields, as shown already by Boltzmann (1876). This is related to the question of bulk viscosity discussed briefly in section IV.

Cauchy initial value problem for the system (1), (23) has a unique solution for "reasonable" initial data. Corresponding theorems (local existence, global uniqueness, and continuous dependence of the solutions on the initial data) have, in fact, been established recently for the collisionless case (see refs. [15a], [15b]), and the general case has essentially also been solved*. These rather deep results show that the kinetic theory model of a gravitating gas is mathematically consistent. The (local) stability of the solutions under small changes of the initial data, combined with Bichteler's result (see [16]) that exponentially bounded initial distributions (i.e., $|f(x,p)| \leq b(x)e^{\beta_a p^a}$ for some b, β_a) remain exponentially bounded for a finite time, lend some credibility to such formal approximation methods as those sketched in section IV.

---

* Private communication from Professor Y. Choquet-Bruhat.

CHAPTER IV. REMARKS ABOUT SPECIAL SOLUTIONS AND APPROXIMATION METHODS FOR NON-EQUILIBRIUM SITUATIONS

a. No exact solutions of the relativistic Boltzmann equation (23), apart from the equilibrium solutions described above, are known if collisions are included (i.e., $W \neq 0$). In the collisionless case, eq. (23) is equivalent to the statement that the distribution function $f(x,p)$ is a first integral of the geodesic equation (5), and since many spacetime models have symmetries which give rise to such first integrals, several solutions of eq. (12) are known. If, e.g., $\xi^a(x)$ is a Killing vector (≡ generator of a one-parameter group of isometries), then the function $\xi_a(x)p^a$ on M is a first integral of eq. (5), whence any positive function of it is a possible collisionless distribution function, and a corresponding remark applies if one has several Killing vectors. (For massless particles, conformal Killing vectors can also be used.) These integrals correspond to the energy, momentum and angular momentum integrals in fields with corresponding symmetries.

The preceding remarks apply in particular to <u>static</u>, <u>spherically symmetric spacetimes</u>, and have been used to compute the general solution of eq. (12) in such spacetimes which is invariant under the full, four dimensional symmetry group* (SO[3]xR). The result can be used to compute T^{ab} - eq. (14) - and to set up the Einstein equation (1). In this way, several solutions of the equations (1), (12) which provide models of <u>relativistic star clusters</u> have been constructed and have been used to estimate the quasistatic evolution of such objects (see references [17],[18]). Also, the <u>stability</u> of such systems against radial perturbations has been studied in a series of

---

* The action of any isometry group of a spacetime X can easily be extended to the phase space M; thus it is meaningful to speak of the invariance of f with respect to such a group.

beautiful papers (references [19], [20]), and the results so far obtained indicate strongly that such clusters become unstable and collapse rapidly as soon as their central redshift exceeds a value of about 0.5, a result which is of interest in connection with a quasar model proposed by Hoyle and Fowler.

Nonstationary solutions of eqs. (1), (12) have been found in connection with cosmological considerations. In particular, it has been established that if a solution has a locally rotationally symmetric distribution function with respect to some mean four-velocity field, then, the mean motion is shear-free and either volume preserving or irrotional; and if it is not volume preserving, the metric must be of the Robertson-Walker type, i.e., it must correspond to a homogeneous and isotropic model universe (refs. [21], [22]). In this case, the first integral on which the distribution function depends is not a linear one associated with a Killing vector - as in the static models - but is quadratic and of the form $(\xi_a \xi_b - \xi_c \xi^c g_{ab}) p^a p^b$, where ξ^a is the conformal Killing vector associated with the isotropic expansion of the universe. (Similar quadratic integrals occur in the corresponding Newtonian solutions, see ref. [23]).

For further applications of kinetic theory to cosmology see references [4], [24], [25], and for some more solutions of eqs. (1), (12) see reference [26].

b. In order to describe non-equilibrium situations one has to resort to approximation methods. Restricting attention to near-equilibrium cases, one can write the actual distribution function f as a "small" perturbation,

$$f = e^{\alpha + \beta_a p^a} (1 + g) = f^{(0)} (1 + g), \qquad (30)$$

of a local equilibrium distribution with parameters $\alpha(x)$, $\beta_a(x)$

whose spacetime variation is to be determined from eq. (23) in conjunction with the small perturbation term g(x,p).

As in nonrelativistic theory one can verify by means of eqs. (22), (13), (14), (18) that the equation of state $\mu = \mu(s,n)$, which relates the equilibrium values of energy density μ, entropy density s and particle density n, remains valid to first order in g for a near-equilibrium distribution (30), if the mean velocity is taken to be $u^a \propto \beta^a$ and μ, s and n are defined, respectively, by eqs. (18), $s = -u_a S^a$, and $n = -u_a N^a$. Similarly one obtains that, to first order in g, the entropy flux relative to the mean motion, $s^a = S^a - su^a$, is related to the diffusion flux $i^a = N^a - nu^a$ and the heat flux q^a (defined through eqs. [18], [19]) by $s^a = \beta q^a - (1 + \alpha)i^a$. Hence, if one matches the parameters α, β^a in (30) to the actual distribution function f by requiring $i^a = 0$, one has the standard thermodynamic relation $s^a = \beta q^a$. Combining these thermodynamic relations with the conservation laws (2) and (25) and using the Gibbs equation

$$d\mu = T\,ds + \frac{\mu + p_0 - s}{n}\,dn \tag{31}$$

to define a temperature T and a thermodynamic pressure p_0, one obtains the expression

$$S^a{}_{;a} = -\frac{1}{T}\{\pi\Theta + \pi_{ab}\sigma^{ab} + q^a(\frac{T_{,a}}{T} + \dot{u}_a)\} \geq 0 \tag{32}$$

for the entropy production rate. Here the kinematical quantities σ_{ab}, Θ and $\dot{u}_a$ are the shear velocity, the expansion rate, and the four-acceleration of the mean flow, defined by

$$\left.\begin{aligned} u_{(a;b)} &= \sigma_{ab} + \tfrac{1}{3}\Theta(g_{ab} + u_a u_b), \\ \sigma_{ab}u^b &= 0, \quad \sigma^a{}_a = 0, \quad \dot{u}_a = u_{a;b}u^b, \end{aligned}\right\} \tag{33}$$

and

$$\pi = p - p_0 \tag{34}$$

is the difference between the total kinetic pressure of eq. (18) and the thermodynamic pressure of eq. (31).

All this follows standard lines of reasoning of nonrelativistic kinetic theory, and shows that the passage from kinetic theory to phenomenological thermo-hydrodynamics can be performed at the relativistic level as easily as in the standard theory, and this also holds for gas mixtures with diffusion and reactions (references [5b] and [32I].

Equation (32) suggests the <u>transport equations</u>

$$\left.\begin{aligned} \pi_{ab} &= 2\eta\,\sigma_{ab}, \\ \pi &= -\xi\Theta, \\ q_a &= -\lambda(\delta_a^b + u_a u^b)\,(T_{,b} + T\,\dot{u}_b), \end{aligned}\right\} \tag{35}$$

with non-negative coefficients, η, ξ, λ. Specifically relativistic terms appear in the heat conduction law only. The acceleration term produces, in an equilibrium state, precisely the temperature variation which has been discussed on page 15 and which is, as we now see, needed to prevent heat from "falling" in a gravitational field, (Equations (35) have been proposed long ago, see ref. [27].)

Instead of guessing equations (35) on the basis of (32) one should, of course, derive them from the Boltzmann equation (23). Two classical methods offer themselves, the <u>Chapman-Enskog method</u> and <u>Grad's method of moments</u>. Both these methods have, in fact, been adapted to relativity; as will be described briefly now.

The Chapman-Enskog method has been adapted to relativity by Israel (reference [6a]) and, in a mathematically more complete form, by Marle (reference [2b]). The method consists of replacing W in equation (23) by $\frac{1}{\varepsilon}W$, expanding g in equation (30) in a power series $g = \sum_{n=1}^{\infty} \varepsilon^n g^{(n)}$, decomposing the Liouville operator (9)

$$\begin{aligned} L = p^a \nabla_a &= \underbrace{-u_b p^b u^a \nabla_a}_{} + \underbrace{(\delta_b^a + u_b u^a) p^b \nabla_a}_{} \\ &= \quad D \quad + \quad \nabla \\ &(\nabla_a = \frac{\partial}{\partial x^a} - \Gamma^{\gamma}_{ab}\, p^b \frac{\partial}{\partial p^\gamma}) \end{aligned} \tag{36}$$

into a "time derivative" D and a "spatial derivative" ∇ (both operating in M), and to solve the resulting equation successively for each power of ε after elimination of $\dot{n}$, $\dot{\mu}$ and $\dot{u}_a$ ($\dot{n} = n_{,a}\, u^a$ etc.) by means of the conservation laws (25), (2). In this procedure the variables n, μ, u_a have to be defined uniquely in terms of the "correct" distribution function f by means of <u>matching conditions</u>, e.g., those of Landau-Lifschitz which require

$$\left.\begin{aligned} \overset{\circ}{T}{}^a{}_b\, u^b &= T^a{}_b\, u^b = -\mu u^a, \\ \overset{\circ}{N}{}^a\, u_a &= N^a\, u_a = -n. \end{aligned}\right\} \tag{37}$$

Here, $\overset{\circ}{N}{}^a$, N^a are the currents (13) formed with $f^{(0)}$, f, respectively, etc. The result of this procedure in first order are the transport equations (35). To obtain the coefficients η, ξ and λ, one has to solve inhomogeneous Fredholm integral equations. This has been done for "Maxwellian particles", defined (in relativity) by having a cross section of the seperable form $\sigma(E,\Theta) \propto (\frac{E}{m})^{-2} ([\frac{E}{2m}]^2 - 1)^{-1/2}\, \Gamma(\Theta)$, and for moderately relativistic temperatures (say, $\frac{T}{m} \lesssim 10^{-2}$), by Israel ($\xi$ and λ, see [6a]) and by de Groot and van Leeuven (ξ,λ and η, see [32,V,VI]). The last-mentioned authors extended these laborious calculations also to non-reacting mixtures of isobaric Maxwellian particles and established the validity of Onsager <u>reciprocity relations</u> for such systems (references [32,I,I,V]). Relativistic corrections to the transport coefficients, all of order $\frac{T}{m}$, have been worked out explicitly. An interesting result of these calculations is that ξ, the <u>bulk viscosity coefficient</u>, is positive, in contrast to the corresponding nonrelativistic result $\xi = 0$ (for a simple gas of point particles). (In Israel's example ξ is independent of n and decreases for low temperature like T^3.) The reason for this deviation of the two theories can easily be seen to lie in the fact that the energy depends differently on the momentum in the two cases; $\xi = 0$ happens to follow from the particular dependence $E = \frac{1}{2m}\vec{p}^{\,2}$ of nonrelativistic mechanics. The nonvanishing of ξ "explains" the descrepancy between these two theories with respect to the existence of expanding local equilibrium flows (p.15): In relativity, even an isotropic expansion is connected with the production of entropy. (For a critical discussion of this point and its bearing on cosmology, see [28].)

Whereas the method just sketched gives only <u>normal solutions</u> of the Boltzmann equation, the method of moments is also capable of describing the "anormal" relaxation towards equilibrium which cannot be described in terms of the "equilibrium variables" n, μ, u_a alone. This method has been modified according to relativistic requirements by Marle [2b] and by Anderson and Stewart (see [3] and volume 9 by Stewart of these "Lecture Notes", [29].)

The basic tool of this method is a <u>complete orthogonal set of functions</u> on the Hilbert space $L^2(P_x, f^{(0)}\pi)$, i.e., the space of real functions on the mass shell P_x which are square integrable with respect to the measure $f^{(0)}\pi$. The set is defined as follows. $\overset{\circ}{H} = 1$, and $\overset{r}{H}{}^{a_1,\dots a_r} = p^{a_1}\dots p^{a_r} - \sum_{s=0}^{r-1} C^{a_1\dots a_r}_{b_1\dots b_s}\overset{s}{H}{}^{b_1\dots b_s}$ $(r = 1,2,\dots)$, with*

$$<\overset{r}{H}{}^{a_1\dots a_r}, \overset{s}{H}{}^{b_1\dots b_s}> = 0 \text{ if } r \neq s. \qquad (38)$$

* The members of the set are the components of the tensors $\overset{r}{H}$. For fixed r, these components are, in general, not orthogonal.

(The C's are constants, and < , > denotes the inner product on the Hilbert space.) Such a set exists, is unique, the tensors $\overset{r}{H}{}^{a_1,\dots a_r}$ are symmetrical and trace free, and the set is complete. These functions form a relativistic analogue of the Hermite-Grad polynomials of R^3, and reduce to them in a suitable limit*. (In contrast to the Hermite-Grad polynomials, the H's cannot be derived from a generating function; they do not obey a Rodrigues relation.)

Assuming that g from eq. (30) is a member of the Hilbert space one can expand it,

$$g(x,p) = \sum_{n=1}^{\infty} \overset{n}{a}_{b_1\dots b_n}(x)\, \overset{n}{H}{}^{b_1\dots b_n}(x,p). \qquad (39)$$

The coefficients $\overset{n}{a}\dots$ can be shown to be linear combinations of moments of f of orders up to n.

Now, the Boltzmann equation (23) implies that

$$\begin{aligned}(\textstyle\int p^{a_1}\dots p^{a_n} f\, \pi)_{;a_n} &= \textstyle\int p^{a_1}\dots p^{a_{n-1}} L(f)\pi \\ &= \textstyle\int p^{a_1}\dots p^{a_{n-1}} (\bar{f}\,\bar{f}' - f\,f')\delta(\Delta P) W \pi\wedge\pi'\wedge\bar{\pi}\wedge\bar{\pi}'\end{aligned} \qquad (40)$$

for n = 1, 2, ..., and conversely this infinite system of equations implies eq. (23). Inserting the expansion (39), the right hand side becomes a quadratic form in the $\overset{n}{a}\dots$ or, equivalently, in the moments of f, with coefficients expressible as integrals involving the $\overset{n}{H}$'s. Hence, (40) represents a system of differential equations for the moments of f which is equivalent to the Boltzmann equation. If one now truncates the series (39) after a few terms and linearises the truncated equations (40) in the $\overset{n}{a}\dots$'s one can obtain a tractable system of partial differential equations for the $\overset{n}{a}\dots$'s (or the moments), and these then define a moment-approximation of eq. (23). Keeping in (39) only the terms with n = 1 and n = 2, one gets the fourteen moment approximation (which corresponds to the

---

* For elegant proofs, see Marle [2b].

nonrelativistic thirteen moment approximation of Grad) which gives just sufficient information to derive again eqs. (35) (and, after Stewart [29], more general equations for gas mixtures with reactions). In addition - and this is one of the principal advantages of this method compared with the first one - explicit integral representations are obtained for the transport coefficients ([3], [29], [6b]). The results concerning ξ given above are confirmed and extended to arbitrary temperatures by this method. Moreover, this method permits to treat general, not only "normal" perturbations, and Stewart has shown [29] that the behaviour of the g-dependent part of T^{ab}, the perturbed part of the stress tensor, is governed by a system of hyperbolic differential equations whose characteristics lie inside the light cone. For a simple Boltzmann gas, the maximal velocity of propagation of such disturbances (relative to the fluid) is $c(\frac{3}{5})^{1/2} \simeq 0.8$ c, which should be compared with the upper limit $\frac{c}{\sqrt{3}} \simeq 0.58$ c for the sound velocity of such a gas [14]. Thus, an old paradox connected with eq. $(35)_3$, the apparently acausal propagation of heat, has been resolved and has been shown to be due to an inadequate approximation. Extensions of this method to relativistic quantum gases are due to Stewart [29] and Israel and Vardalas [30].

As a last remark I wish to mention that a method which treats photons or neutrinos as a "gas" described by a distribution function, and which describes the medium with which this radiation interacts as a fluid - an approximation which is useful in astrophysical problems - has been worked out in general relativity by Lindquist [31]; several applications of this theory of radiative transfer have been made, and more work along these lines is being carried out.

In conclusion it may be said that the basic conceptual and formal framework of relativistic kinetic theory is now well established, and that this new branch of statistical physics has proven to be a valuable tool of research which offers many possibilities for further investigation.

References:

[1] N. A. CHERNIKOV: Acta Phys. Polon. 23, 629 (1963); 26, 1069 (1964), and earlier papers cited therein

[2] C. MARLE: (a) Ann. Inst. Henri Poincaré A 10, 67 (1969) (b) Ann. Inst. Henri Poincaré A 10, 127 (1969)

[3] J. L. ANDERSON: in Relativity, ed. M. Carmeli, S. J. Fickler, L. Witten (London: Plenum Press, 1970), p. 109

[4] R. K. SACHS AND J. EHLERS: in Astrophysics and General Relativity, ed. M. Chrétien, S. Deser and J. Goldstein (New York: Gordon and Breach, 1971) Vol. 2, p. 331.

[5] J. EHLERS: (a) contribution to Proceedings of the International School of Physics "Enrico Fermi" Course 47 (New York: Academic Press, 1971), p. 1
(b) contribution to Relativistic Fluid Dynamics (Roma: Edizioni Cremonese, 1971), p. 301

[6] W. ISRAEL: (a) Journ. Math. Phys. 4, 1163 (1963)
(b) contribution to Studies in Relativity (Oxford: Clarendon Press, to be published in 1972)

[7] F. JÜTTNER: (a) Ann. Phys. 34, 856 (1911) (b) Ann. Phys. 35, 145 (1911) (c) Z. Physik 47, 542 (1928)

[8] J. L. SYNGE: Trans. Roy. Soc. Canada III 28, 127 (1934)

[9] A. G. WALKER: Proc. Edinburgh Math. Soc. 4, 238 (1936)

[10] A. LICHNEROWICZ AND R. MARROT: Comp. Rend. Acad. Sci. (France) 210, 759 (1940)

[11] G. E. TAUBER AND J. W. WEINBERG: Phys. Rev. 122, 1342 (1961)

[12] J. EHLERS: Abh. Akad. Wiss. Mainz (Jahrg. 1961), 791

[13] J. L. SYNGE: Relativity: The Special Theory (Amsterdam: North-Holland Publishing Co., 1956)

[14] J. L. SYNGE: The Relativistic Gas (Amsterdam: North-Holland Publishing Co., 1957)

[15] Y. CHOQUET-BRUHAT: (a) Journ. Math. Phys. 11, 3228 (1970) (b) Ann. de l'Institut Fourier (to appear in 1971)

[16] K. BICHTELER: Commun. Math Phys. 4, 352 (1967)

[17] YA. B. ZEL'DOVICH AND M. A. PODURETS: Astr. Zh. 42, 963 (1965); english translation in Soviet Astron. AJ 9, 742 (1966)

[18] E. D. FACKERELL: (a) Ap. J. 153, 643 (1968) (b) Ap. J. 165, 489 (1971)

[19] J. R. IPSER AND K. S. THORNE: Ap. J. 154, 251 (1968)

[20] J. R. IPSER: (a) Ap. J. 156, 509 (1969) (b) Ap. J. 158, 17 (1969)

[21] J. EHLERS, P. GEREN AND R. K. SACHS: Journ. Math. Phys. 9, 1344 (1968)

[22] R. TRECIOKAS AND G. F. R. ELLIS: "Isotropic Solutions of the Einstein-Boltzmann Equations" preprint Univ. of Campridge, 1971.

[23] J. EHLERS AND W. RIENSTRA: Ap. J. 155, 105 (1969)

[24] C. W. MISNER: Ap. J. 151, 431 (1968)

[25] R. A. MATZNER: Ap. J. 157, 1085 (1969)

[26] R. BEREZDIVIN AND R. K. SACHS: in Relativity ed. M. Carmeli, S. J. Fickler and L. Witten (London: Plenum Press, 1970), p. 125.

[27] C. ECKART: Phys. Rev. 58, 919 (1940)

[28] E. L. SCHÜCKING AND E. A. SPIEGEL: Comments Astrophys. Space Physics 2, 121 (1970)

[29] J. L. STEWART: Non-Equilibrium Relativistic Kinetic Theory , Lecture Notes in Physics (Berlin: Springer-Verlag, 1971), volume 9

[30] W. ISRAEL AND J. N. VARDALAS: Nuovo Cimento Ser. I, 4, 887 (1970)

[31] R. W. LINDQUIST: Annals of Physics 37, 487 (1966)

[32] S. R. DE GROOT, C. G. VAN WEERT, W. TH. HERMENS, AND W. A. VAN LEEUWEN: Physica 40, 257 (1968); 40, 581 (1969); 42, 309 (1969); W. A. Van Leeuwen and S. R. de Groot, Physica 51, 1; 16; 32 (1971).

COMPUTER EXPERIMENTS ON SELF-GRAVITATING SYSTEMS

Richard Miller
University of Chicago
Chicago, Illinois

CHAPTER I. INTRODUCTION

Gravitational n-body calculations usually come in two rather different forms. The usual kind of n-body calculation is the obvious kind: consider n point masses in space, compute the force on each as exerted by all the remaining (n-1) point masses under an inverse square law of forces, and then integrate the Newtonian equations of motion as accurately as you can to find the motion of the n bodies. Since this kind of calculation necessarily requires computing the force for each particle pair, the amount of computation required increases drastically with increasing numbers of particles. A practical upper limit to the number of particles attained in calculations of this kind is about 500.

The second kind of gravitational n-body calculation is based on the use of smoothed particle-aggregates. These can take any of a variety of forms--spherical shells for spherically symmetrical systems, or just some kind of "blob" that moves rigidly in cartesian coordinates, for example. The advantage of this formulation is that various approximations can be used to obtain the forces, sidestepping the need for detailed summing of the forces between all particle pairs. This permits the treatment of many more "particles,"--up to 10^5 without any great difficulty.

In Chapter II, we consider the application of a calculation of the second kind to an astronomical problem--the formation and persistence of spiral structure in spiral galaxies. The calculation used runs in cartesian coordinates, using a four-dimensional phase space corresponding to the motion of point particles on a plane in configuration space. Because the astronomical context of this problem may be unfamiliar, a considerable part of the discussion is devoted to description of spiral galaxies and to pointing out the astronomical basis of the formulation as well as the kinds of things sought in the computational results.*

Two numerical experiments which involve the more traditional approach to n-body calculations are discussed in Chapter III. In these experiments the dynamics of a small (32-body) stellar system is followed "exactly." First, we examine the role of pair correlations in stellar systems, suggesting a calculable two-body variable which provides an estimate of the pair correlation. In the second part we consider some of the difficulties encountered in attempts to perform "thermodynamic experiments" on self-gravitating systems.

---

*Work of this kind has been done by two groups: that of Roger Hockney and Frank Hohl at NASA Langley, and the group at the Institute for Space Studies in New York with Kevin Prendergast and William Quirk of Columbia University, with whom I have been associated. The emphasis in Chapter II will naturally be on the work with which I am most familiar, that with which I have been associated.

CHAPTER II. SPIRAL GALAXIES IN A COMPUTER

Spiral structure is a puzzle of long standing in astronomy. Spirals cannot be a transitory evolutionary phase; the statistics of relative numbers of spirals among all galaxies (about 2/3 spirals) are not consistent with such a notion. The basic problem is how to keep the spirals from wrapping themselves up. They are known to rotate (from spectroscopic evidence), and do not rotate rigidly. The typical field of differential rotations has larger angular velocities (even linear velocities) in the inner portions than there are farther out. Any pattern impressed on such a differentially-rotating form would wrap up and become indistinguishable after a few rotations. And the rotation times (around 1/4 billion years for our Galaxy) would wash out any spiral patterns in times far too short to be consistent with large fraction of spiral galaxies actually observed. This loss of detail in the pattern is illustrated in a sequence of steps of the wrapping-up process in Figure 1.

The way out of this difficulty was given by B. Lindblad, who started working on this problem around 1925. The "wrapping-up" occurs if the spiral consists of identifiable material--what, today we call a "material arm." Lindblad pictured the spiral as a pattern--a density wave. C. C. Lin and his group have built this idea into a pleasing theory that has caused quite a bit of excitement among astronomers.

The density wave is familiar to nearly everyone who has watched a traffic jam on an expressway at rush hour. Cars approaching the jam from behind find their way blocked by the large batch of cars in the jam--they thus attach themselves to the rear edge of the traffic jam. But cars at the front edge find essentially a clear road ahead, and so can leave the traffic jam and proceed into the clear road. If you come upon a traffic jam, you will notice that the cars around you at the time you enter remain around you as you proceed through the

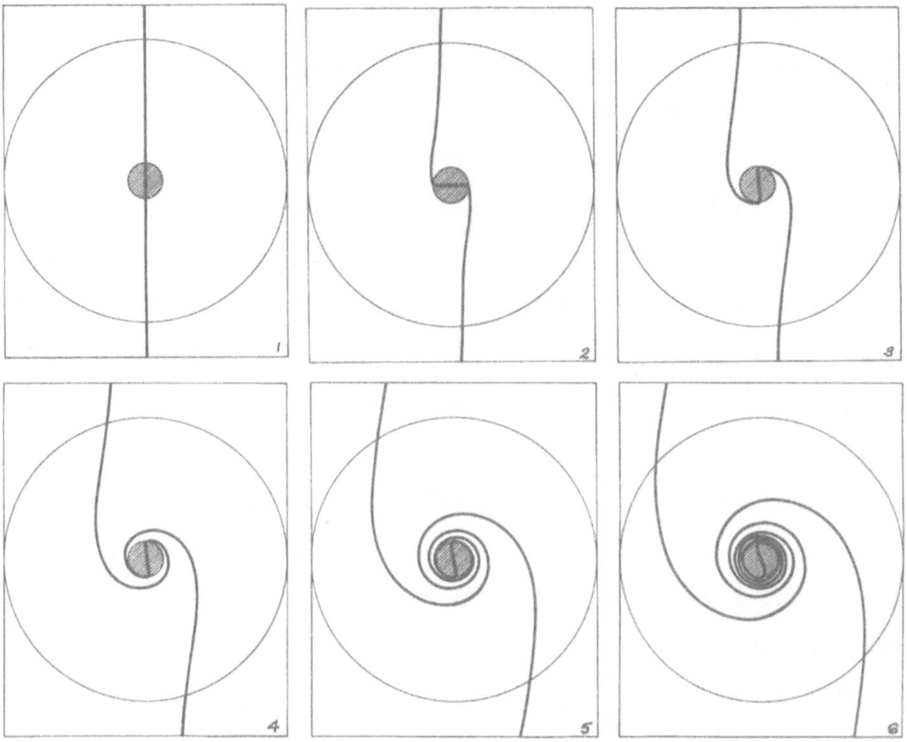

FIGURE 1. Six stages in the wrapping-up of a pattern (in this case a straight line) that follows a field of differential rotations. The crowding of the pattern near the center makes the spiral pattern nearly indistinguishable after about 3-5 rotations. For most galaxies, this would "wash out" the spiral patterns in about 1 billion years.

traffic jam. But after a while, you work your way through the traffic jam, and become one of the cars on the front edge, ready to proceed into the clear road and detach yourself from the traffic jam. The cars that were near you as you entered are still near you. Both you and the traffic crest have moved forward along the expressway--but you have moved faster than the traffic jam. The traffic jam is a density wave. Of course, in the galaxy problem, the situation is a little bit different. The "expressway" is turned into a closed loop, and the problem must be done in a rotating coordinate system. We usually use a coordinate system in which the traffic jam is at rest--but the rotating coordinate system means that the dynamics is to be done as if the motion were being studied on a merry-go-round.

Even with the help of these models, the problems of spiral structure cannot be regarded as solved--many features are not yet understood. Neither theory nor observation can give an unambiguous answer to so simple a question as whether the spiral patterns lead or trail. The lifetime and stability of the spiral patterns are open questions. There are other difficulties as well, but they merely reinforce the need for alternate approaches to the problem.

One of the nice things about working in astronomy is that many of the objects are incredibly beautiful. A spiral galaxy is one of the most pleasing objects. In Figure 2, a well-known spiral galaxy, M 51 is shown. Ignore the bright knot at the end of one of the arms. The features that impress you immediately in this photograph are (1) a general rather good twofold symmetry that extends over the entire galaxy, in spite of many detailed irregularities, (2) a reasonable amount of contrast, or of brightness difference, between the spiral arms and the interarm regions, (3) generally decreasing brightness farther from the center, with a fairly bright center (the photograph does not show this nearly as well as it should---no photograph can), (4) disappearance of the spiral pattern at the center, but the spiral continues outward as far as you can distinguish the galaxy, (5) some dark lanes on the inner edges of the spiral arms (trailing edges, if the arms trail), (6) it looks like a flat object seen face-on in this photograph--it is a little difficult to imagine a three-dimensional structure that would look something like this in any direction, (7) it is clearly a self-gravitating system, (8) while there may be neighbors their influence is small. The experts will see a lot more in this picture. The bright spots outside the galaxy image are foreground stars.

FIGURE 2. A spiral galaxy seen face-on. This is known to astronomers as M51 or NGC 5194.

Figure 2 was made with an ordinary (i.e., blue-sensitive) photographic plate. If a red-sensitive plate is used behind a filter to remove most of the blue light, the galaxy shows much less structure. If the galaxy is photographed through a filter that passes light of the Balmer series of Hydrogen (H_{α} or H_{β}), then a set of bright "knots" is seen along the spiral arms--principally near the dark lanes. The interpretation of this is that most of the light in the spiral arms comes from very bright, young, blue stars (stars known to astronomers as O and B stars), and from ionized hydrogen

(HII) regions surrounding such stars. These stars, which may be as much as 1000 to 10000 times as bright as the sun, but only 10-30 times as massive, do not live very long--they consume the available fuel stores much too rapidly. The red background may come from stars that are less massive, hence longer-lived. Most of the mass is in the form of stars that produce the red light, most of the light comes from the blue stars.

Our own Galaxy has all these ingredients as well. We see bright blue stars, many faint red stars, and gas clouds. There are some bright red stars too, but these are much less massive than the bright blue stars, and are at an advanced stage of their aging process. Usually, the bright blue stars are near or inside gas clouds, the gas very near the star often being ionized. The gas density is very irregular. The bright blue stars must have been born recently--presumably out of the concentrations of gas. Moving with typical velocities, they cannot depart from their parental gas cloud very far during their lifetime. Of course, stars of all masses will be formed from these gas clouds--many more low-mass stars than high-mass stars, but almost all the light comes from the bright blue (massive) stars. It is, of course, no accident that we think that extragalactic nebulae are built of the same kinds of objects that we see in the solar neighborhood of our own Galaxy--it is precisely because we see them here that we think they must be the principal constituents of other galaxies. We also see in our own Galaxy dark regions, or "dust clouds," usually associated with gas clouds and bright blue stars, that we think are similar to the dark lanes in these other galaxies.

NGC 1300, in Figure 3, shows another common form of spiral galaxy. This is known as a "barred spiral", and shows the same features as have been pointed out in M51, although perhaps different

FIGURE 3. A different kind of spiral galaxy--a barred spiral. This is NGC 1300, also seen face-on.

in detail. The barred spirals usually have the pair of dark lanes symmetrically disposed near the ends of the bar. The bar tends to be redder than the arms. M81, in Figure 4, is a particularly beautiful object, showing again the same kinds of features. Here the spiral pattern is more tightly wound. M81 gives the distinct impression of a flat object seen in some direction other than face-on.

FIGURE 4. Another spiral galaxy (M 81 = NGC 3031). This gives the impression of a flat object seen from some angle other than face-on.

Finally, NGC 891, in Figure 5, shows the extreme case of one of these objects seen edge-on. Presumably, if you could see it from another direction, NGC 891, might look like M 81 or M 51 (without the satellite). Notice the dark lane concentrated rather closely to the median plane. All these pictures are shown in an attempt to convince you that a reasonable model for these objects is a self-gravitating mixture of various constituents, all constrained to move in a plane. These are not a sequence of photographs of the same object as seen from various directions. That is a luxury astronomers do not have - we cannot go around and look at our objects from the other side.

FIGURE 5. A spiral galaxy seen edge-on (NGC 891). Presumably M 51 or M 81 would look like this if viewed from the appropriate direction, and NGC 891 might look like one of them if viewed from another direction.

There are other kinds of galaxies--principally the very regular and beautiful ellipticals, which look like (oblate) ellipsoidal mass distributions, and do not show the dark lanes or gaseous regions--and the irregulars, which show a little bit of everything, with much less organization.

Nothing has been said about magnetic fields. A few years ago, most attempts to explain spiral structure centered on magnetic fields. There is good evidence that magnetic fields are one of the ingredients of our galaxy. The main justification for omitting them from the present discussion is that the influence of the magnetic fields on the dynamics of the stars is through the gravitational effect of the ionized gas--which represents a small fraction of the total mass. Failure to construct a convincing spiral model without magnetic fields would force us to include them; but it is worth a try without magnetic fields because a model without them will be much simpler.

The starting point for most current theories of spiral structure is abstracted from the conditions just described. Models are to be constructed of self-gravitating systems restricted to a plane. In that plane, there is a predominantly axisymmetric mass distribution that generates axisymmetric potential and force fields. The axisymmetric part consists of red stars and contains most of the mass. Superimposed on this background is a gaseous system--also self-gravitating, but obeying gas-dynamical equations rather than the particle equations of the stellar dynamical system. The two sub-systems partake of the differential rotation. The gaseous subsystem contains a spiral pattern which rotates (almost) rigidly with its own angular velocity. The material (both stars and gas) flows through the pattern. There is a slight potential minimum at the pattern (the total potential field is no longer axisymmetric), where the gas tends to concentrate. The gas concentration also induces a slight concentration of stars in the neighborhood of the spiral pattern, but that concentration is much weaker. A shock may form as the gas flow enters the potential minimum at the spiral pattern. New stars are thought to form in regions of high density--thus preferentially near the shock. When old stars die, they return some gas to the medium, to allow this process to continue. However, not all gas is returned, so the process cannot go on forever. The angular velocity of the pattern is lower than the angular velocity of the gas and stars over most of the region in which the pattern can be seen. Lin's models are built by impressing a spiral pattern on this kind of background, then solving the self-consistency problem for the combination of gas and stars in the (linearized) limit of small density variations and of small pitch angles for the spiral patterns.

Computer models, on the other hand, may start from nearly axisymmetric models and allow a process like star formation to go on. The stars move under the usual stellar dynamical equations, with the forces determined by self-gravitation. The "gas" population follows a modified dynamics according to which turbulent energy is artificially removed. So far, in our calculations, the "gas" has not obeyed gas-dynamical equations, but only a crude approximation to them. A shock could not form in these models. We are improving this feature of the calculation. Hohl's models differ in imposing an axisymmetric potential in which the stars move. Computer models normally handle about 10^5 particles--they could be pushed to 10^6 or 10^7 on current machines if there were any clear-cut reason for doing so. Even so, they fall far short of the 10^{11} in a real galaxy. Thus the theoretical models (Lin, and others) and the computer models are complementary approximations to real stellar systems. The theoretical models ignore the grainy structure of real stellar systems, while the computer models are far too grainy.

Details of the calculations have been published, and will not be discussed here. Our calculation has been advertised to be reversible and to have an exact Liouville theorem in the μ-space, all obtained at the cost of treating the integrations somewhat crudely. Reversibility is as much a matter of numerical accuracy and roundoff as it is of the difference-scheme used. We have taken some pains in these matters, but cannot give an honest appraisal as to how important these features are.

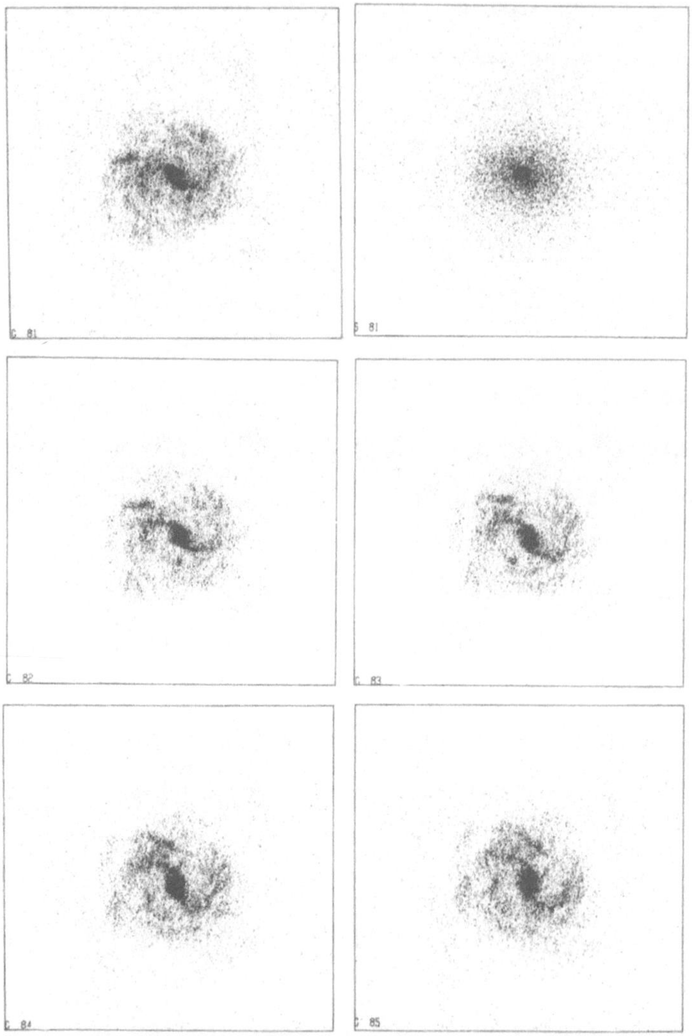

FIGURE 6. Frames from the motion picture of the computer spirals. The upper-right-hand frame shows the "stars", which change little during the calculation; the rest show the "gas" at various integration steps.

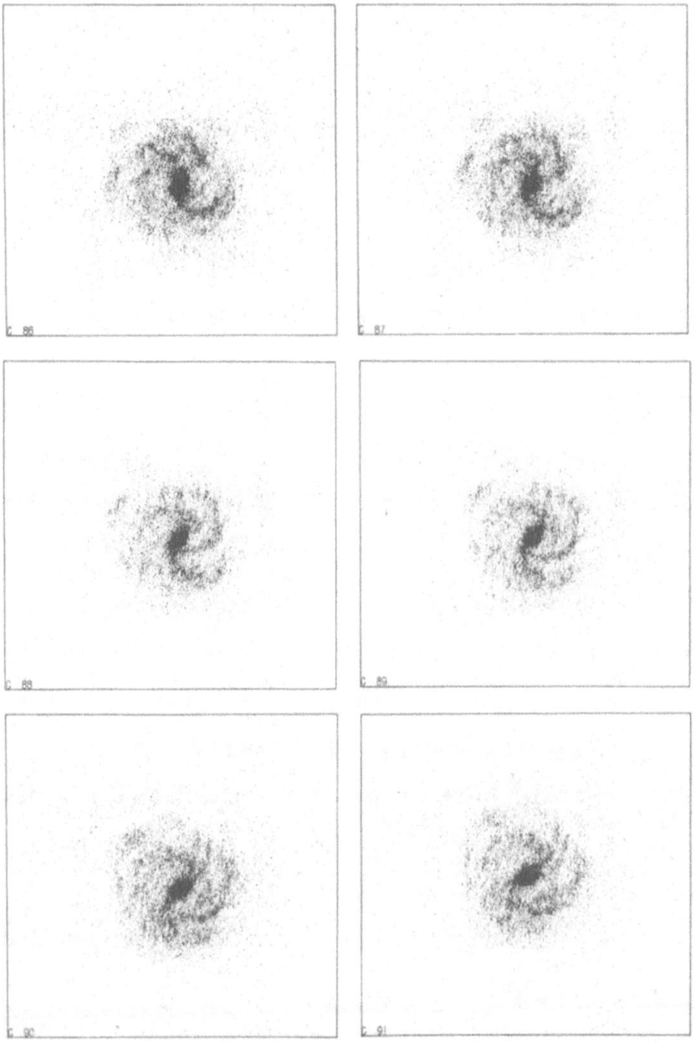

FIGURE 7. Continuation of Figure 6.

The results of a calculation that yielded spiral patterns are shown in a motion picture. A few frames from the motion picture are reproduced here as Figures 6 and 7. The "star" field shown in the upper right-hand corner of Figure 6 changes very little as the calculation proceeds. The remainder of Figures 6 and 7 show the "gas" at successive integration steps at a stage of the calculation in which the spiral pattern had settled down fairly well. The pattern rotates in about 30 integration steps, while Figures 6 and 7 show 11.

This calculation started from a circular disk that was all "gas", but had a rule for creating "stars" out of the "gas" that is thought to conform to what might go on in a real galaxy. The precise rule must have profound dynamical consequences--certainly altering the rule alters our models--but this particular aspect of the calculation should not be taken too literally. By the time shown in Figures 6 and 7, about 85% of the mass was in the form of "stars", the remaining 15% still being "gas". Star formation had stopped long before the time of these figures. However, once stars were formed, they remained stars for the rest of the calculation--there was none of the recycling of material that is expected in a real stellar system.

The spiral density wave idea is shown in Figures 8 and 9. These figures represent the "gas" portion of the system, with a few "particles" singled out and plotted as large squares. The identity of certain "particles" is retained from frame to frame. In Figures 8 and 9, individual "particles" can be seen to approach the spiral feature from behind (the rotation is clockwise), dwell at the feature momentarily, then to pass on through it. We have not been able to show this effect in a sequence of still pictures nearly as dramatically as the motion picture shows it, but the effect is there.

A word of warning. These sequences--and the motion picture--should not be considered as depicting the aging or evolution of a real galaxy. The initial conditions are certainly unrealistic, and the

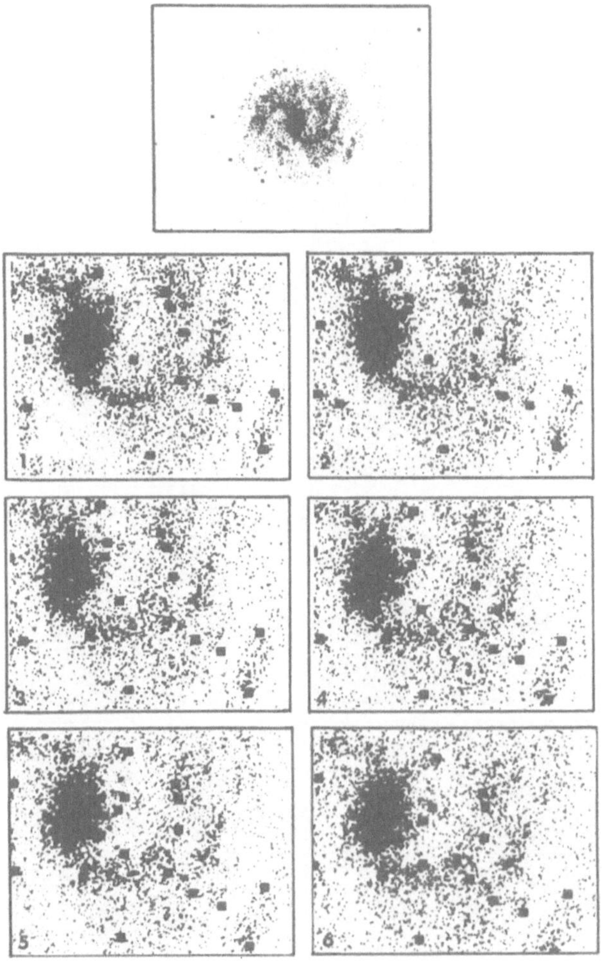

FIGURE 8. Details of the spiral patterns, showing individual particles moving through the spiral features. The entire system is shown in the top frame, the lower frames are enlargements out of that picture at intervals of 1/5 an integration step. Certain particles are plotted as large squares in each of the frames, they show the motion of those particles relative to the pattern.

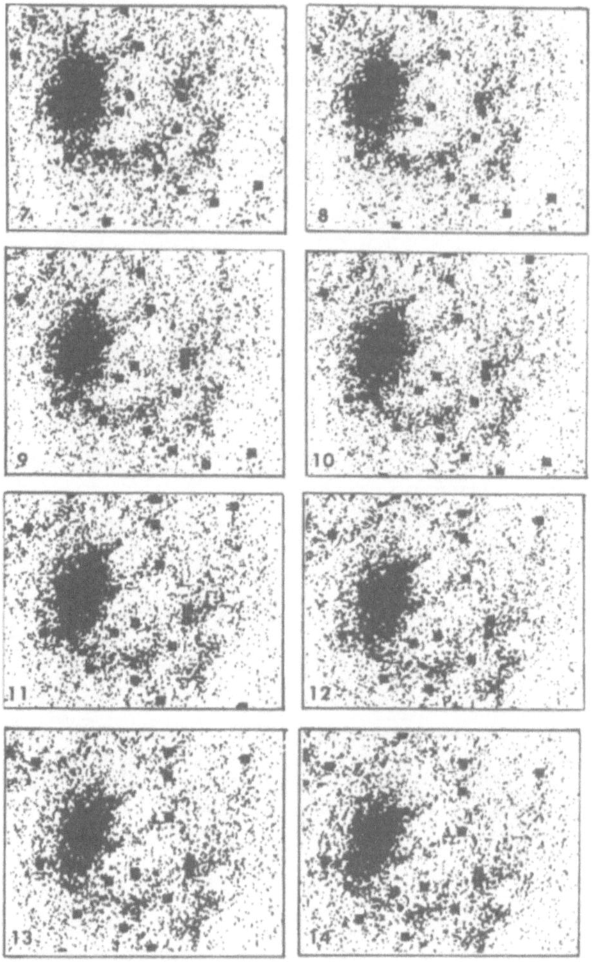

FIGURE 9. Continuation of Figure 8.

real system has a more complex interplay of various properties than we have been able to include in the computer program. Thus the fact that one or more of these frames might look like photographs of some real galaxy does not mean that we have managed to reproduce the evolutionary history of that particular galaxy, or even that earlier and later stages of the calculation indicate what the galaxy would look like at earlier or later stages of its aging process. The value of the numerical experiments lies in general indications of how difficult it is to build spiral patterns that live for a while (these lived for about 3-4 pattern rotations), of the interplay between the "star" and "gas" populations in the pattern, what fraction of the mass of the system participates in the pattern, and so on. It is particularly valuable to be able to "kick" the computer experiment to see if it "bounces",--something that we cannot do with the real galaxy. The computer experiment is an experimental tool with which we can try to find out what makes spiral patterns. While the emphasis in this paper is on spiral patterns, there are other experiments that both Hohl and we have done with these systems. These include experiments on gravitational stability and attempts to verify various stability predictions.

So what have we learned from these computer experiments? We have learned that spiral patterns can be constructed of self-gravitating systems without need to invoke magnetic forces. But two populations were needed, or some other artifice to emphasize the spiral pattern. Real galaxies have a very effective amplifier of small density variations to produce large brightness variations. Computer models need the same thing. On detailed analysis, we find that there is a spiral density wave in the stars as well as in the gas--there is about as much total mass participating in the spiral phenomenon in the stars as there is in the gas. But about 1/4 of the gas participates and less than 5% of the stars, so when we plot the star density we do not see a spiral pattern. So far, there are no spiral patterns in computer models without two populations (think of Hohls' background potential as the second population), just as we know of no spiral systems in the sky that do not have two populations. But are two populations necessary? We do not know. Spiral patterns seem to appear when the conditions are about right, but we find that spiral patterns are difficult to stir up if the conditions are not just right.

As with all experiments, the computer experimenter must be very careful to avoid interpreting situations in which the experimental results fail to contradict his prior prejudices as proof of the correctness of those prejudices. With these experiments, we have seen some patterns. What we see fails to contradict our prejudices. We feel that we have a foot in the door, and a valuable tool for experimenting with properties of spiral systems. The real test comes now--to see if we know how to use that tool for some definitive experiments.

References:

For those who may wish to pursue some of these matters further, the following references are starting-points and lead to earlier literature:

[1] Theoretical work on spiral waves, especially the Lin school:
C. C. LIN, C. YUAN, AND F. H. SHU: Astrophys. Journ. 155, 721 (1969).
A particularly readable account of Lin's theory was given by G. Contopoulos. pp. 303-16 of the conference report referred to in (4) below.

[2] Hohl's computer experiments:
F. HOHL: "Dynamical Evolution of Disk Galaxies," Technical Report NASA TR R-343, July 1970.
F. HOHL AND R. HOCKNEY: Journ. Computational Physics 4, 306 (1969).

[3] Our group:
R. H. MILLER, K. H. PRENDERGAST, AND W. J. QUIRK: Astrophys. Journ. 161, 903 (1970).
R. H. MILLER: Journal of Computational Physics 6, 449-72 (1970).

[4] General review of astronomical view of spirals (conference proceedings)
The Spiral Structure of Our Galaxy, in IAU, Symposium 38
W. BECKER AND G. CONTOPOULOS, EDITORS. (Dordrecht, Holland: D. Reidel Publishing Co.) 1970.

CHAPTER III. NUMERICAL EXPERIMENTS ON PAIR CORRELATIONS AND ON "THERMODYNAMICS"

A "conventional" n-body calculation was constructed to carry out a number of numerical experiments on small stellar systems. Two of these experiments will be reported here. These experiments refer to a 32-body system, in which the equations of motion were handled "exactly", with the usual checks on constancy of the ten first integrals of motion. Because the emphasis was on the experiments, the calculation was constructed without particular attention to running speed or minimum storage requirements.

These calculations were quite different in spirit and in the formulation from the rather crude models with very many particles that led to the spiral patterns described in Chapter II.

The computer programs were conventional n-body routines, like those described some years ago by von Hoerner (1960, 1963). Most of the special methods now commonly used to speed up the calculation (Aarseth, 1963, 1966; Wielen 1967, 1968) were not incorporated; fast-running calculations were not as important for this project as ready adaptability to special purposes. An iterative refinement feature optionally permitted the first ten integrals of motion to be refined to arbitrary precision. Comparison runs were made with and without this feature.

Units were chosen so $G = 1$, $m = 1$ (for each particle). All particles have the same mass; systems with a variety of particle masses typically evolve rather rapidly to a state in which the most massive particles have sought one another out to form tight binaries (Aarseth, 1968). The additional complications of different masses would make this set of experiments quite unmanageable.

A. Pair Correlations

The reduced distribution functions provide a useful way of visualizing stellar dynamical problems, but are difficult to apply to actual stellar systems because they are too general--there are not many theoretical reasons for restricting the wide class of admissible functions and very little observational or experimental evidence. The situation is bad enough with the single-particle functions, but is substantially worse with regard to two-particle or pair terms. The pair term is particularly interesting because the fundamental interaction is between particle pairs.

It is convenient to separate the pair correlation term in the two-particle reduced distribution function

$$f_2 = f_1 \cdot f_1 + g \qquad (1)$$

Observationally or experimentally, evidence on g is difficult to obtain, because it is constructed from $(f_2 - f_1 f_1)$ and involves taking the difference of large quantities, each of which is difficult to determine accurately.

Some time ago, a study was made to see what could be learned about g from catalogued observational data concerning stars in the solar neighborhood (Miller, 1967). While stars seem to be closer together in configuration space than they would without g, it was apparent that good determinations would require much more accurate observational data than there is any hope of obtaining. An attractive alternative seems to be provided by numerical experiments with n-body systems.

One of the principal difficulties facing this effort to demonstrate the character of g is to find some property of the system that behaves quite differently with nonzero g from the way in which it would behave if g were zero. A consequence is that most tests that can be devised to demonstrate the existence of a nonzero g are weak.

Effects due to nonuniform f_1 can easily be misinterpreted as representing a pair correlation. For example, a test based on the magnitude of the vector between the two particles in configuration space would have the effects of the pair correlation intertwined with f_1 in a very complex way.

1. Pairing in Phase Space

The most convincing demonstration of the existence (and importance) of g follows from the observation that interacting particles cannot occupy the same phase point in the μ-space. Noninteracting particles may, of course. In the limiting case f_2 should go to zero as the two phase points approach each other. Think of what it means for two particles to be near the same point in μ-space. They are close together in configuration space, and there is no relative velocity. If there is an interaction--particularly, an interaction that becomes strong at short range, there must be a rather large force between them, and that force will rather quickly develop a relative velocity. Thus, the two particles will no longer be near each other in μ-space.

This effect is shown in Figure 1. Particle pairs are represented by points plotted according to the magnitude of their separations in configuration (abscissa) and velocity (ordinate) space. Points tend to avoid regions of small r_{12} and v_{12}, as might be expected, since there is little available volume there. But they also avoid a sensibly larger region in which both are small.

This region may be related to the energy of pair interaction, or the binding energy that a particle pair would have in the absence of all other particles:

$$E_{pr} = \frac{1}{2} (\underset{\sim}{v}_1 - \underset{\sim}{v}_2)^2 - \frac{2}{(\underset{\sim}{x}_1 - \underset{\sim}{x}_2)} \quad . \tag{2}$$

Note that E_{pr} is not a conserved quantity. The uppermost curve in Figure 2 has $E_{pr} = 0$, the lower curve is for circular orbits. Note the way in which the density of points diminishes in the region where $E_{pr} < 0$, with very few pairs in that region. The case plotted in Figure 1 started with 16 pairs on the curve representing circular orbits at $r_{12} = 0.5$; these pairs moved out of the region of bound pairs long before the time represented.

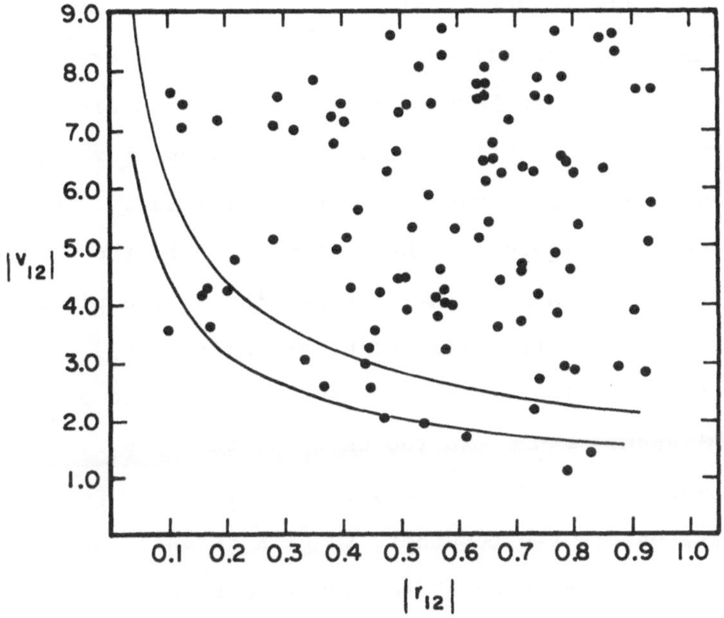

FIGURE 1. Evidence for pair correlations in the avoidance of the same phase point in the μ-space. Each plotted point represents one particle pair, plotted according to $r_{12} = |x_1 - x_2|$ and $v_{12} = |\tilde{v}_1 - \tilde{v}_2|$. Points avoid the region where both $\tilde{r}_{12}$ and $\tilde{v}_{12}$ are small.

Singling out E_{pr} as the dominant variable is a matter of prejudice; functions of r_{12} and v_{12} other than E_{pr} can give similarly shaped equidensity contours in Figure 1. The effect shown in Figure 1 is not just a phase-volume effect, as becomes evident when it is considered that permuting the v_{12}'s associated with the r_{12}'s would destroy the correlation in the pair phase space and the effect described without affecting the projections to pair configuration space and to pair velocity space.

Additional evidence favoring the selection of E_{pr} as the dominant variable is given in Figure 2. There, the logarithm of the number of pairs with E_{pr} greater than some selected value is plotted as a function of the selected value for a number of experiments. While there is some tendency for the histograms to curve toward negative E_{pr} for small numbers of particle pairs, a clear inhibition is present. This is particularly shown for the cases in which the pair correlation was destroyed artificially: comparison of Plots D and C of Figure 2. Above about n = 15, the dominance of f_1 terms is evident. A straight-line portion in the plots of Figure 2 would represent a distribution going as $\exp(+E_{pr}/E_o)$, for some value E_o to be determined from the sample. For these plots, E_o takes on a value of about 1 - 2 in the units of this paper.

The two-body reduced distribution function, f_2, goes to zero when the phase points are together, so g is strongly negative there. The pair correlation g is equal to minus the square of f_1 along the ray through the 12-dimensional pair phase space in which both members of the pair are in the same point of the μ-space.

The same effect shows up in other ways as well. For example, the pair that is closest in configuration space is very nearly the most widely separated in velocity space. It may or may not have negative E_{pr}. The pairs that are closest in velocity space tend to be well separated in configuration space, but they do seem

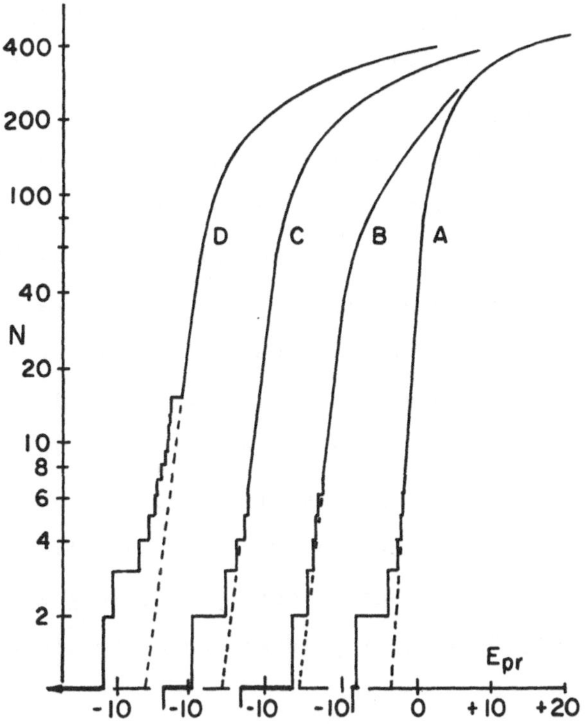

FIGURE 2. Histograms of log N vs. E_{pr}, where N is the cumulative number of pairs of particles whose E_{pr} is less than the value of E_{pr} appearing on the abscissa. The individual plots are displaced by 10 units of E_{pr} along the abscissa for clarity. The dashed lines are straight-line extrapolations to N = 1 of the linear portion of each histogram. Plot C is the experiment referred to in Figure 1. Plot B represents the same experiment at a slightly earlier time (0.3 crossing-times instead of 0.4). Plot D was obtained from the data of Plot C by permuting the r_{ij} associated with a given v_{ij}. The value of E_{pr} for N = 1 is -126 for Plot D. Note that this plot is quite similar to Plot C above N = 15, but tends toward more negative E_{pr} for N<15. The most spectacular difference is at N = 1, of course. Plot A is a different experiment (different initial conditions) for comparison purposes, run to 1.5 crossing-times.

systematically to have negative E_{pr}. The history of the example plotted in Figure 1 is typical. The closest pair in configuration space was farthest apart in velocity space; it had strongly negative energy ($E_{pr} = -9.8$), but the system was just recovering from a very close collision involving that pair. At short times (1/10 crossing time) before and after the time of Figure 1, that pair had positive E_{pr}, the value of E_{pr} being nearly the mean for all pairs.

Thus, there is a very strong g operating in these systems. This comes as no surprise--the gravitational interaction is strong enough at short range that there should be no two stars at the same phase point. It is characteristic of this effect due to g that it shows up very strongly in a few pairs, and weakly if at all in the systematic statistics of the entire set. The strong appearance in a few easily identified pairs simplifies the experimental problem of studying the effect.

2. Discussion

Some positive regions have been found for g, and a strongly negative region at small phase separations, implying zeroes in between (with due regard for dependence on 8 variables!). But even such simple things as the locations of zeroes are not very well determined. There must be regions of both signs in order to preserve the normalization of both f_1 and f_2. Inspection of Figure 1 suggests that there is a zero of g near a surface $E_{pr} = 0$. This surface has the kind of hyperbolic shape shown in Figure 1 when viewed along the ray through the pair phase space in which the two particles have the same μ-space coordinates. The function g projects to the pair configuration space to be weakly positive near the ray in which the two particles have the same configuration space coordinates; the projection to pair configuration space presumably gets more positive as the particles get closer and closer in configuration space. The projection onto the pair velocity space may behave similarly, although the effect is much weaker.

The function g must go to zero at infinity in order that both f_1 and f_2 may do the same. The regularity conditions commonly applied to distribution functions must apply to g as well.

The vanishing of f_2 when its arguments coincide must carry on into the higher-order reduced distribution functions as well. This means that g must be of the same order (in terms of particle numbers) as f_2. Think of f_1 and f_2 as being normalized to unity. Along the ray of the 12-dimensional space in which the μ-space coordinates of the two particles are identical, $f_2 = 0$, so $g = -f_1 f_1 = -f_1^2$. Now, imagine placing more particles in the system, while retaining the normalization on the f's. The statement about $g = -f_1^2$ still holds, and since f_1 is independent of particle number, g must be also. Clearly the same argument applies right on up through the hierarchy to the n-particle correlation function defined analogously to g.

The tendency for g to go positive in projections to the pair configuration space was noted earlier in the catalogued data of stars in the solar neighborhood (Miller, 1967). Stars in observed systems also strikingly avoid the same point of μ-space. But observed systems have a strong positive g for negative E_{pr} that manifests itself in the form of the observed binary (and multiple) systems. The zero of g is moved to strongly negative E_{pr}. There may be considerable additional structure to g for negative E_{pr} in real stellar systems. Evidently binaries in observed systems were born binary--systems in numerical experiments do not develop nearly as many close binaries. But the persistence of binaries in real systems shows that the dynamics causes slow changes (if any) in g where E_{pr} is strongly negative.

These matters have been discussed more fully in a published paper (Miller, 1971).

It should be stressed that this effect on g is a consequence of the presence of an interaction between particle pairs, and does not depend upon the property of the interaction's being inverse-square in this investigation. The conclusion is of much broader generality.

3. Some Speculations

The avoidance of the same place in the μ-space is reminiscent of the exclusion principle in quantum mechanics. It is inviting to speculate that the exclusion noted here, resulting from the interaction between particle pairs, might lead to a carving up of the μ-space into cells, just as the Planck constant does in quantum mechanics. One of the more troublesome difficulties in classical statistical mechanics is that there is no natural unit in which to measure volumes in the μ-space. Perhaps the pairing of interacting particles could accomplish the same effect.

A difficulty remaining for this program is that there is no obvious unit in which to measure the cell size, even in this problem. However, the pair energy, E_{pr} has the property that the phase volume (the volume in μ-space) available to a particle pair with $E_{pr} < \alpha$, say, is finite (it is $[-\alpha]^{-3/2}$).

Crudely, one might say that the value of α appropriate to a cluster would be determined from E_o, the slope of the distribution of pair energies determined from Figure 2. An alternative would be to partition the total cluster energy (which must be negative for a bound cluster) among the available pairs, noting that the sum of all the pair energies (over all pairs) is

$$\Sigma E_{pr} = n\, T_{tot} + 2\Omega = (n-2)T_{tot} + 2E_{tot} \qquad (3)$$

and ΣE_{pr} has a least value of twice the total cluster energy. Either way, the phase volume argument is likely to be restricted to finite clusters--for infinite systems, the total energy is not defined, and the slope, E_o, is expected to steepen with more particles in the cluster. The restriction to finite systems for the self-gravitating case is not surprising.

I believe that the carving up of the phase space by the pair correlation is essentially correct; I would like very much to find a way to make the arguments more convincing. The exclusion implies that the proper statistics of stellar systems is something like a Fermi statistics--although in practice, it must lead to a high temperature limit of Fermi statistics and thus to a statistics that is not distinguishable from the usual Boltzmann statistics.

An additional feature of the exclusion of regions in the μ-space corresponding to two particles too close together comes in the application to the usual discussions of relaxation times due to binary encounters. Most physics students have probably seen the derivation of the multiple scattering formula for charged particles in matter in terms of the transverse impulse delivered to a particle travelling on a straight line. The resulting integral has a logarithmic divergence at both near and distant limits. The same treatment leads to the same trouble in estimating the relaxation time due to binary encounters in stellar dynamics. Chandrasekhar (1960) has shown how the divergence at close encounters can be removed by properly accounting for the details of the orbit--it is a Keplerian hyperbolic orbit if one ignores the effects of all other particles. That calculation proceeds from the assumption $E_{pr}>0$, in the language of this note. A remarkable feature is that the divergence at close encounters can also be removed even in the straight-line orbits, if one says that the least value of E_{pr} along the orbit should be zero. This, of course, has an _ad hoc_ flavor, but then almost all treatments of binary encounters have an _ad hoc_ flavor if they are investigated in sufficient detail. But the conclusion remains that it is the prohibition of negative E_{pr} that has succeeded in removing the divergence at close encounters in the relaxation time calculation.

In this sense the notion of avoiding negative E_{pr} is familiar to most astronomers.

B. "Thermodynamics"

An n-body calculation provides one way of attempting the thermodynamic "gedankenexperimente" of using a stellar system as the thermodynamic medium in a Carnot engine. These experiments have not gotten as far as putting the system into a Carnot engine, although routines have been constructed to do that. The work to be reported was undertaken to explore the practical problems that might be encountered. How long would a system have to sit to reach an equilibrium state? How fast could a cylinder wall move and still carry the system through a sequence of quasiequilibrium states? Or is an equilibrium state attainable at all? The interaction of a stellar system with a "thermodynamic enclosure"--even one without moving walls--can provide a good starting point.

We will not here enter into a discussion of the applicability of thermodynamic concepts to self-gravitating systems. Self-gravitating systems fail to satisfy even the most basic axioms of thermodynamics. But still, thermodynamic methods are of such power and generality as to make it seem worthwhile at least to try to apply some of the concepts to stellar systems. As you will see, the failure to behave thermodynamically, even in the most rudimentary way, is the principal result of this investigation. Even the most basic operation, that of trying to define a thermodynamic temperature by some operational process, fails. This section is a chronicle of what happens when you blindly try some "thermodynamic experiments" on self-gravitating systems.

The system was placed in a box. Whenever a star struck the box, it was caught and thrown back into the box. It was re-projected randomly directed inward, with a randomly selected speed such that its new kinetic energy would be exponentially distributed with a mean value that was called the "temperature" of the box. The star was re-projected from the same point, so there was no change of potential energy. Changes of momentum and of kinetic energy were tallied. Finally, the box was endowed with a "heat capacity", such that the difference of kinetic energy received and given up could cause a change in the "temperature" of the box. The exponential distribution of kinetic energy is appropriate to a Maxwellian velocity distribution. The random velocities of re-projection are satisfactory for all parts of the Carnot cycle, although a specular reflection might be easier to apply to the adiabatic part.

Two sequences of experiments were run. For the first sequence, the box was "cold"--its temperature was always zero. The second sequence used non-zero temperatures and various heat capacities. The final equilibrium state of the second sequence should define a thermodynamic temperature for the stellar system by the classical definition of temperatures from the equilibrium of two systems in contact, if indeed such an equilibrium can be reached. It is not clear that an equilibrium can be reached, even by ascribing a negative heat capacity to the box.

The system should quickly readjust itself to something it likes from any initial condition. In fact, initially the particles randomly filled a rectangular parallelopiped in the phase space whose projection onto the configuration space and onto the momentum space was a cube in each space. The cube edge in configuration space was adjustable, but always within the box, while the cube edge in momentum space was chosen to make the ratio of kinetic to potential energy correspond to the virial theorem.

The cold box was typical of most experiments. Since there is an infinite energy store available, arbitrarily large amounts of energy can be transferred to the enclosure. The history of energy transfer to the enclosure in the longest run is shown in Figure 3 (solid curve;

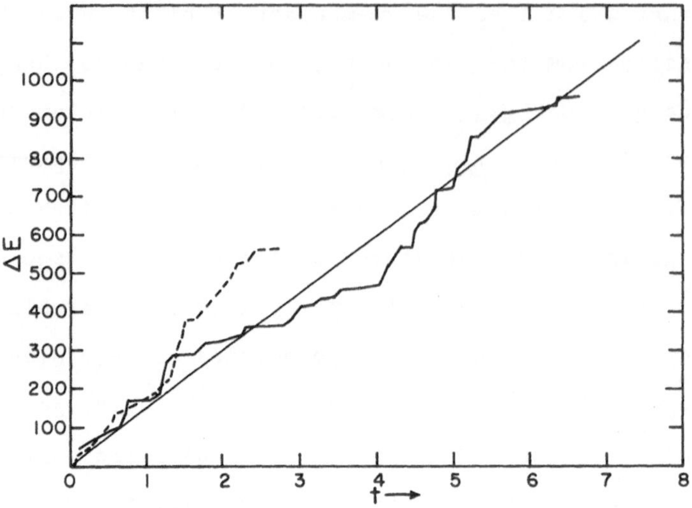

FIGURE 3. Energy transfer from a 32-body system to a cold enclosure. The total initial kinetic energy was about 250. The time units on the abscissa are nearly a crossing time of the initial state. The two tracks represent two distinct calculations.

the dashed curve shows a shorter run with a different starting condition to illustrate the reproducibility of the experiments). The rate of energy transfer to the enclosure is nearly constant over the duration of the experiment--it does not diminish or accelerate appreciably over the duration of the run--even though the total energy transferred to the enclosure was about 4 times the initial kinetic energy present in the system. (The initial kinetic energy was about 250 in the units of the ordinate). It was characteristic of these runs that the virial theorem was approximately satisfied without allowing for "surface pressure" terms. Thus, near the end of the run, the stellar system was very small and compact, somewhere within the box, with very high kinetic energy. The time-scale in Figure 3 is the time-variable of the differential equations, which is the time-scale of an observer looking at the cluster from outside. In time-units appropriate to the cluster (crossing-times, for example: a crossing-time is the time it would take a particle having the mean kinetic energy of particles in the system to travel a distance equal to some characteristic dimension of the cluster), the rate of energy transfer decreases. The crossing-time (estimated from the virial theorem) was about 1 time-unit at the beginning of the calculation, but was about 1/10 time-unit at the end.

There is no preference for any one star to fly out--the identity of stars that strike the box seems random. Occasionally, it seems that the same one comes out repeatedly, but the effect is never significant. At the end of the long run, the selection of stars that have struck the enclosure seems to be drawn from the appropriate multinomial distribution.

There was very little change in the average amount of energy carried by a star that struck the enclosure during the calculation; a possible slight trend toward increasing the amount of energy carried is heavily masked in the fluctuations. The average energy delivered to the box by a particle striking the wall was about the same as the average kinetic energy of one particle in the initial condition. Similarly, the rate at which stars struck the enclosure did not show a detectable trend. The distribution of intervals between times when stars struck the box seemed random--it looked like a Poisson process with a fixed rate at about 10-12 per initial crossing time. There was no detectable serial correlation of intervals.

The zero-temperature case is interesting because the stars flying out are caught, their kinetic energy removed, and they are merely released where caught. They start falling toward the center of the remaining cluster--and have very little chance to avoid passing through the most dense part. This arrangement is very efficient at extracting energy from the cluster. It is almost a Maxwell demon.

The rate of energy transfer to the box was about half the total initial kinetic energy in the system per initial crossing time.

The second series of experiments, that with nonzero box temperature, did not tend toward an equilibrium (Figure 4). It tended to have the star system give up energy to the box. With the box temperature colder than the stellar system, this is what would be expected. But with the box hotter than the system, the system still gave up energy to the box. This is not in the direction expected of a system with a definable "specific heat." (See the discussion by Lynden-Bell and Wood, 1968, with which a good deal of this work can be compared. The discussion of Thirring, 1970, also applies.) The general trends are: (1) With the box temperature up to about that of the average star, the energy transfer to the box occurred at nearly the rate observed with a cold box. (2) With the box temperature about twice that of the average star, the rate was appreciably diminished. (3) With the hotter box, the fluctuations in the energy transferred were quite large--fluctuations as great as half the initial kinetic energy of the system were not unusual, although we were unable to provide convincing evidence that the large amplitude fluctuations represent a cooperative effect. The fluctuations in the lower three plots of Figure 4 are noticeably larger than those in the uppermost plot. A general trend of transferring energy to the box was superposed on the large fluctuations.

Qualitatively, the results of these experiments were anticipated. It was expected that fluctuations in the complexion of a stellar system would lead to localities that were hot, and in which a "heat imbalance" would exist compared to the rest of the system. The questions to be answered by experiment had to do with whether these fluctuations would last long enough to lead us to a locally unstable situation. The usual continuum theories (see, e.g., Lynden-Bell and Wood, 1968, Larson, 1970) tend to minimize or to ignore these

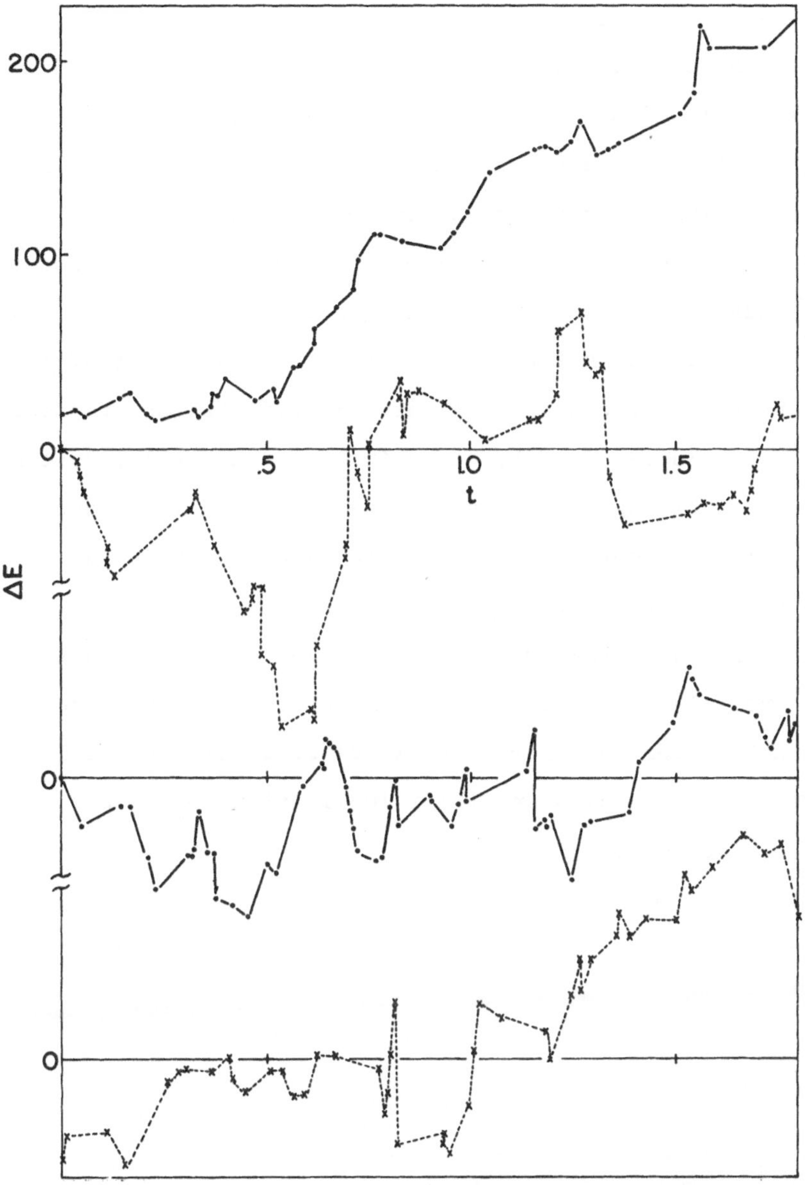

FIGURE 4. Energy transfer from a 32 particle system to an enclosure with non-zero temperature. The track at the top resulted with the enclosure temperature equal to the mean kinetic energy of a particle in the initial state. The three lower tracks (to the same scale on the ordinate, but displaced for clarity) represent three different calculations in which the enclosure temperature was twice the initial mean kinetic energy per particle. All systems later ran away. The large fluctuations cannot convincingly be shown to represent cooperative phenomena.

fluctuations. (Formally, they are there, hidden in some of the higher moments of the distribution functions; when the sequence of moment equations is closed by relating high-order moments to lower order moments, the fluctuations are ignored at that order of the moment equations.) Fluctuations should become less important as the number of particles increases, but the amplifying effect of the localized "runaway" makes fluctuations have a much larger effect than expected--their importance does not diminish like the square root of the particle number, for example. A naive interpretation of these experiments would lead to the conjecture that a stellar system might always be unstable against such "runaways." The importance of localized fluctuations does not seem to have been stressed in discussions applied to stellar systems.

Similarly, with the cold box, the unanticipated result is quantitative. The rate of energy transfer to the box is surprising. It goes much faster than Larson's (1970) discussion would lead you to expect. These results and their implications will be discussed more fully elsewhere.

ACKNOWLEDGEMENTS

The work leading to Chapter II was greatly assisted by Professor K. H. Prendergast of Columbia University, who has been deeply involved in all phases of this work from its inception, and by Dr. William J. Quirk, of Hale Observatories, who joined the project shortly after it started. The bulk of the computations have been carried out at the Goddard Institute for Space Studies in New York, through the courtesy of Dr. Robert Jastrow, Director. The work has been assisted by the U. S. Atomic Energy Commission. Part of the work was done with the aid of a National Research Council Senior Postdoctoral Resident Research Associateship at the Goddard Institute for Space Studies, supported by the National Aeronautics and Space Administration.

It is a pleasure to thank J. Hofslund for his help with the experimental results, and Professors P. O. Vandervoort and I. Lerche for many helpful discussions on the material of Chapter III. This work was started at the Kitt Peak National Observatory, while the writer was there as Consulting Astronomer. The use of the computational facilities at Kitt Peak is gratefully acknowledged. Support from the Shirley Farr Fund of the University of Chicago was very helpful. This work was also partially supported by the U. S. Atomic Energy Commission.

References:

[1] S. J. AARSETH: Mon. Not. Royal Astron. Soc. 126, 223 (1963)

[2] S J. AARSETH: Mon. Not. Royal Astron. Soc. 132, 35 (1966).

[3] S. J. AARSETH: Bull. Astronomique,3,No. 3, 105 (1968).

[4] S. CHANDRASEKHAR: Principles of Stellar Dynamics, Chapter II, (Dover, 1960).

[5] S. V. HOERNER: Zeitscr. F. Astrophysik 50, 184 (1960).

[6] S. V. HOERNER: Zeitscr. F. Astrophysik 57, 47 (1963).

[7] R. B. LARSON: Mon. Not. Royal Astron. Soc. 147,323-37 (1970).

[8] D. LYNDEN-BELL AND R. WOOD: Mon. Not. Royal Astron. Soc. 138, 495-525 (1968).

[9] R. H. MILLER: Astrophys. Journ. 148, 865 (1967).

[10] R. H. MILLER: Astrophys. Journ. 165, 391-7 (1971).

[11] W. THIRRING: Zeit. Physik 235, 339-52 (1970).

[12] R. WIELEN: Ver. Astron. Rechen-Instituts Heidelberg Nr. 19 (1967).

[13] R. WIELEN: Bull. Astronomique,3, 3, 127 (1968).

PROPAGATION OF WAVES IN DISCRETE MEDIA, HARMONIC, ANHARMONIC, AND DEFECTIVE

Elliott W. Montroll
The University of Rochester
Rochester, New York

FOREWORD

The propagation of sound, shock, and detonation waves in continuous media has been investigated in great detail, but theoretical information is sparce on wave propagation in discrete media in the non-linear regime and in the linear regime in the presence of irregularities. The aim of this lecture is to summarize the difference between the propagation of small amplitude waves in continuous and discrete media and to present some of the features of non-linear wave propagation in discrete media, and of both linear and non-linear wave propagation in irregular discrete media.

CHAPTER I. COMPARISON OF PROPAGATION OF SMALL AMPLITUDE WAVES IN CONTINUOUS AND ONE-DIMENSIONAL DISCRETE MEDIA

The wave equation for the propagation of small amplitude waves in a continuous medium (without an energy dissipation mechanism) is

$$f_{tt} = c^2 f_{xx} \tag{1.1}$$

(where $f_x \equiv \partial f/\partial x$, etc., c the propagation velocity and f the physical quantity, for example density, whose variation in the medium is determined by the wave propagation). Any function

$$f = f(x \pm ct) \tag{1.2}$$

is a solution of (1.1). Hence, any initial disturbance of f from a constant value retains its form while propagating with velocity c either to the left or to the right. Even a very sharp gradient in an initial disturbance retains its shape as it propagates.

A medium composed of springs and masses generally does not allow an arbitrary wave to propagate without a change in form. The one-dimensional discrete analogue of (1.1) is the equation

$$d^2 f_n/dt^2 = (\gamma/m)[f_{n+1} - 2f_n + f_{n-1}] \tag{1.3}$$

where f_n is the displacement of the nth mass from its equilibrium position. If both sides of (1.3) are divided by a^2 (a being the lattice spacing), the resulting equation

$$\frac{m}{\gamma a^2}\frac{d^2 f_n}{dt^2} = \frac{1}{a}\left\{\frac{f_{n+1}-f_n}{a} - \frac{f_n - f_{n-1}}{a}\right\} \tag{1.4a}$$

reduces to (1.1) as $a \to 0$ if one defines

$$c^{-2} = \lim_{\substack{a\to 0\\ m/\gamma\to 0}} m/\gamma a^2 \quad \text{and} \quad (f_{n+1} - f_n)/a = \partial f/\partial x. \tag{1.4b}$$

A solution of (1.3) is

$$f_n = \exp i(n\theta \pm \omega t) \tag{1.5}$$

if θ and ω are related by the dispersion relation

$$\omega^2 = (2\gamma/m)(1 - \cos\theta) \tag{1.6a}$$

or

$$\omega = \omega_L \sin(\theta/2) \quad \text{if} \quad \omega_L = (4\gamma/m)^{1/2}. \tag{1.6b}$$

The appropriate values of θ (and, therefore, of the normal mode frequencies ω), depend on the end conditions on the chain. For example, if the chain is composed of N particles and is closed in the form of a ring, then $f_{n+N} \equiv f_n$ so that one must have

$$\theta = 2\pi k/N \quad \text{where} \quad k = 1,2,3,\ldots,N. \tag{1.7}$$

The wave forms

$$\cos(n\theta \pm \omega t) \quad \text{and} \quad \sin(n\theta \pm \omega t),$$

which are obtained by taking linear combinations of (1.5) and its complex conjugates yield waves which propagate to the left (for + sign) or to the right (for - sign) with velocity (a being the lattice spacing)

$$\begin{aligned} v &= a\, dn/dt = \pm a\omega/\theta \\ &= \pm \frac{Na\omega_L \sin(\pi k/N)}{2\pi k} \approx (\tfrac{1}{2}a\omega_L)\{1 - \tfrac{2}{3}\pi^2 k^2 a^2/L^2 + \ldots\}, \end{aligned} \tag{1.8}$$

which depends on k with $L \equiv aN$ being the length of the chain. In the long wavelength regime, $ka \ll L$, v is constant as in the continuum case.

Since an arbitrary initial condition can be expressed as a linear combination of normal modes of (1.5), we find that any wave with a sharp wave front, or any wave which is restricted to a limited region of space small compared with the total chain length, contains many k components. The small k components will travel with velocity $\frac{1}{2}a\omega_L$ independently of k, while the large k components will travel with a velocity of 0(N/k). Hence the wave form will lose some of its sharpness and spread because of the dispersive effect of a discrete medium.[1]

A convenient way of following the development of a wave form as it progresses through a chain is to express the solution of the equation of motion (1.3) in terms of Bessel functions as was first emphasized by Schrodinger[2]. Let

$$u_{2n} = \dot{f}_n \quad \text{and} \quad u_{2n+1} = \tfrac{1}{2}\omega_L(f_n - f_{n+1}) . \tag{1.9a}$$

Then the equations of motion (1.3) in terms of the u_n's become

$$\frac{2du_n}{d\tau} = (u_{n-1} - u_{n+1}) \quad \text{with} \quad \tau = t\omega_L . \tag{1.9b}$$

This equation is exactly the same form as the Bessel function equation

$$J_{n-1}(\tau) - J_{n+1}(\tau) = 2dJ_n(\tau)/d\tau . \tag{1.10a}$$

Another useful Bessel function equation is the recurrence formula

$$J_{n-1}(\tau) + J_{n+1}(\tau) = (2n/\tau)J_n(\tau). \tag{1.10b}$$

We see then that u_n can be expressed as a linear combination of Bessel functions

$$J_{n-r}(t\omega_L) \qquad r = 0, \pm 1, \pm 2, \ldots \tag{1.11}$$

If initially all displacements f_n are zero and all velocities except the mth are initially zero, then

$$u_n(t) = \dot{f}_m(0)J_{n-2m}(\tau) \tag{1.12}$$

is a solution of (1.9b). On the other hand, if velocities are initially zero and the mth displacement is initially the only non-vanishing one,

$$u_{2m+1}(0) = \tfrac{1}{2}\omega_L f_m(0) \qquad \text{and} \qquad u_{2m-1}(0) = -\tfrac{1}{2}\omega_L f_m(0), \quad (1.13)$$

so that

$$\begin{aligned} u_n(t) &= \tfrac{1}{2}\omega_L f_m(0)\{J_{n-2m-1}(\tau) - J_{n-2m+1}(\tau)\} \\ &= \omega_L f_m(0) dJ_{n-2m}/d\tau . \end{aligned} \quad (1.14)$$

Since our original equations (1.9) are linear, a superposition of these special solutions (1.12) and (1.14) leads to

$$\begin{aligned} u_n(t) &= \sum_{m=-\infty}^{\infty} \{\dot{f}_m(0)J_{n-2m}(\tau)+\omega_L f_m(0)\ dJ_{n-2m}/d\tau\} \\ &= \sum_{m=-\infty}^{\infty} \{\dot{f}_m(0)J_{n-2m}(\tau) + \tfrac{1}{2}\omega_L[f_m(0)-f_{m+1}(0)]J_{n-2m-1}(\tau)\}. \end{aligned} \quad (1.15)$$

The particle velocities and displacements are then, respectively,

$$\dot{f}_n(t) = u_{2n}(t) \quad (1.16a)$$

and

$$f_n(t) = f_n(0) + \int_0^t u_{2n}(t)\ dt . \quad (1.16b)$$

Some interesting special cases[3] are (a) particle at origin displaced by unity at t = 0, all other particles at equilibrium positions, and all particles initially at rest

$$f_n(t) = J_{2n}(t\omega_L) \qquad \text{if} \qquad \dot{f}_n(0) = 0 \qquad \text{and} \qquad f_n(0) = \delta_{n,0}; \quad (1.17)$$

and (b) particle at origin at equilibrium but moving with velocity unity, all others fixed at equilibrium position

$$f_n(t) = \int_0^t J_{2n}(t\omega_L)\ dt \text{ if } \dot{f}_n(0) = \delta_{n,0} \quad \text{and} \quad f_n(0) = 0 \quad (1.18)$$

so that

$$f_n(t) = \frac{2}{\omega_L} \sum_{\nu=0}^{\infty} J_{2n+2\nu+1}(t\omega_L). \tag{1.19}$$

In the case of a semi-infinite chain, the particle "-1" is uncoupled with "0" so that the end condition of the chain is

$$\frac{d^2 f_o}{dt^2} = \frac{\gamma}{m}(f_o - f_1), \tag{1.20}$$

or, in terms of the u's

$$2du_o/d\tau = -u_1. \tag{1.21}$$

It is easy to verify that if $f_o(0) = 1$ and all else vanishes at $t = 0$,

$$f_n(t) = J_{2n}(t\omega_L) + J_{2n+2}(t\omega_L) \quad \text{for } n = 0, 1, 2, \ldots \tag{1.22}$$

while, if $\dot{f}_o(0) = 1$, and all else vanishes at $t = 0$,

$$f_n(t) = \int_o^t dt\{J_{2n}(t\omega_L) + J_{2n+2}(t\omega_L)\} \quad \text{for } n = 0, 1, 2, \ldots \tag{1.23}$$

The case in which the end particle, $n = 0$, is driven to the right with a constant velocity u, with all other particles initially at rest at their equilibrium positions, is a simple model of a shock produced by a driving plate on a bar. The solution of our equation of motion is

$$f_n(t) = \frac{2u}{\omega_L} \sum_{\nu=0}^{\infty} (2\nu+1) J_{2n+2\nu+1}(t\omega_L). \tag{1.24}$$

We have plotted the case of an initial impulse (1.23) and of a driven first particle in Fig. 1a and 1b. In a non-dispersive medium, all particles would have the same trajectory as $f_o(t)$ but lagging it with a time depending on the particle number. However, as is clear from the figure 1, this is not the case for our chain of particles.

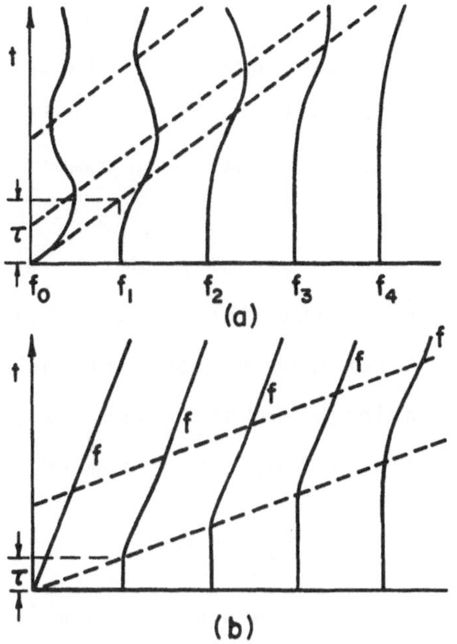

FIGURE 1. Curves (a) represent the trajectories of various masses when the end particle is given an initial impulse. Curves (b) represent the situation in which the end particle is driven with a constant velocity[3].

In general, the wave form broadens and flattens as it propagates and is accompanied by a diminishing train of ripples, both in space and time. It was pointed out by Gilman and Vineyard[4] that the pulse in (1.17) reaches the nth particle at time

$$t \simeq \frac{2n}{\omega_L} + \frac{0.809(2n)^{1/3}}{\omega_L} \quad \text{if } n >> 1.$$

Neither the linear continuous or discrete models exhibit shock wave character. There is no dependence of the wave shape on amplitude and the characteristic sharpening of the wave form does not occur.

The simplest one-dimensional chain with nonharmonic interactions in which shock waves develop is a chain of hard spheres which interact only through their mutual infinite repulsion at short distances. We discuss[4] this case now. Certain other more complicated models will be presented later.

Consider a one-dimensional gas of hard spheres which are initially equally spaced and at rest so that the distance between the centers of successive spheres is a. Let m be the mass of each sphere.

The laws of conservation of momentum and energy between two colliding spheres of mass m_1 and m_2 can be summed up in the equations

$$m_1u_1 + m_2u_2 = m_1v \tag{1.25a}$$

$$\tfrac{1}{2}m_1u_1^2 + \tfrac{1}{2}m_2u_2^2 = \tfrac{1}{2}m_1v^2 , \tag{1.25b}$$

where u_j is the velocity after collision of the sphere of mass m_j and it is postulated that before collision, that of mass m_2 is at rest while that of mass m_1 has a velocity v. It is easily seen that

$$u_2 = \frac{2v}{1+(m_2/m_1)} \begin{cases} \longrightarrow v \text{ as } m_1 \to m_2 \\ \longrightarrow 2v \text{ as } m_1 \to \infty \end{cases} \tag{1.26a}$$

$$u_1 = \frac{v(1-m_2/m_1)}{1+(m_2/m_1)} \begin{cases} \longrightarrow 0 \text{ as } m_1 \to m_2 \\ \longrightarrow v \text{ as } m_1 \to \infty \end{cases} \tag{1.26b}$$

These formulae can now be applied to the propagation of a disturbance down the line of equal masses which we discussed above to be originally at rest and uniformly spaced. Let us first analyze the case in which the particle at the left end of the chain moves to the right with velocity v. After it collides with the first particle to its right, we see from the upper arrow cases of eq. (1.26a) and (1.26b) that the incident particle stops, giving its velocity to the one with which it collides, which then achieves the velocity v. This continues down the line. If the diameter of each sphere is Δ, the time between collisions is

$$t = (a-\Delta)/v \quad . \tag{1.27}$$

The mean speed at which the disturbance propagates is the ratio of distance it proceeds in each collision to the time required for that propagation,

$$V = a/[(a-\Delta)/v] = v/[1-(\Delta/a)] > v. \tag{1.28}$$

After the distrubance passes. the atoms are again at rest but they have moved a distance $(a-\Delta)$ to the right.

A case which is more interesting is that in which the particle at the left end is driven with a constant velocity v which, through a continually applied driving force, is not effected by collisions. This is equivalent to giving the left end particle an infinite mass and corresponds to the lower arrow case of equations (1.26a) and (1.26b). Hence the velocity of the leading edge of the disturbance is, (replacing v in (1.28) by 2v),

$$V = 2v/[1-\Delta/a]. \tag{1.29}$$

Since the velocity of the trailing edge is v, the average thickness of the pulse is

$$\delta = \left\{\frac{1+(\Delta/a)}{1-(\Delta/a)}\right\} vt \tag{1.30}$$

and v is the velocity of particles in the shocked region. The trajectories of both of the cases discussed are plotted[4] in figures (2) and (3).

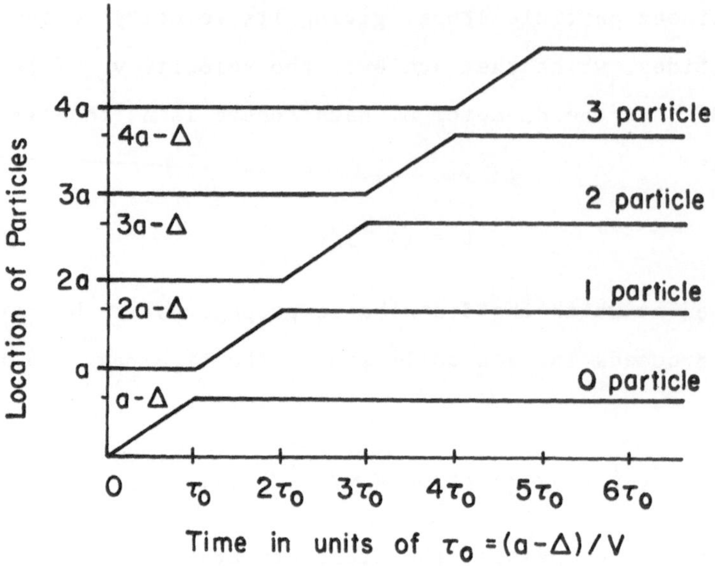

FIGURE 2. Trajectories of a line of hard spheres when the left most particle has an initial velocity V and the rest of the spheres are initially stationery.

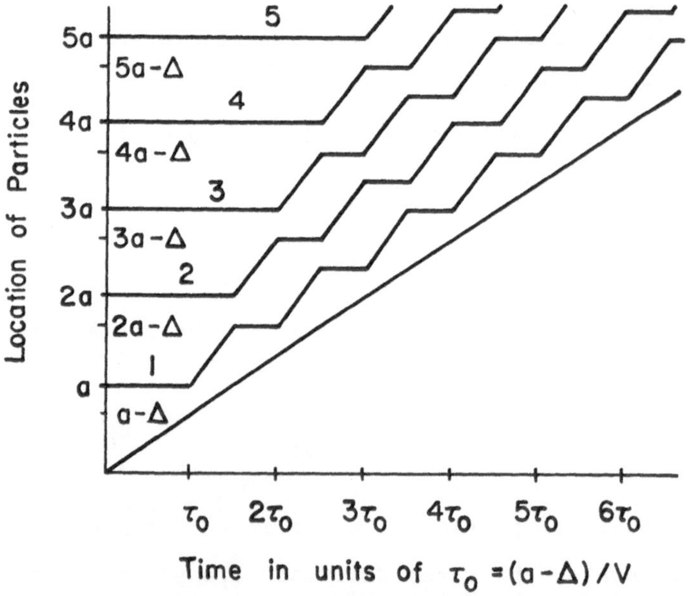

FIGURE 3. Trajectories of a line of hard spheres when the left sphere is driven with a constant velocity V.

CHAPTER II. EQUATIONS OF MOTION AND DISPERSION RELATIONS IN 2D AND 3D HARMONIC LATTICES[5]

We now examine the two-and three-dimensional models which correspond more closely to real solids. The equations of motion of a crystal lattice follow immediately from the crystal Hamiltonian which has the form

$$H = T + \Phi \quad ; \tag{2.1a}$$

T being the kinetic energy

$$T = \tfrac{1}{2} \Sigma\, m_k\, \dot{u}^2_\alpha\binom{\ell}{\kappa} \quad , \tag{2.1b}$$

and Φ the potential energy

$$\Phi = \Phi_0 + \tfrac{1}{2} \Sigma\, \Phi_{\alpha\beta}\binom{\ell\ \ \ell'}{\kappa\ \ \kappa'} u_\alpha\binom{\ell}{\kappa} u_\beta\binom{\ell'}{\kappa'} + 0(u^3) \tag{2.1c}$$

where, in the harmonic approximation terms, cubic and higher order in displacements from equilibrium are neglected.

The quantities

$$u_\alpha\binom{\ell}{\kappa} \qquad \text{and} \qquad \Phi_{\alpha\beta}\binom{\ell\ \ \ell'}{\kappa\ \ \kappa'}$$

represent respectively the displacement from equilibrium of the particle (of mass m_κ) located at the κth position in the ℓth unit cell (α running through the set of components, x, y, and z) and the force constant which couples the displacement in the α direction of the κth atom in the ℓth cell with that in the α' direction of the κ' atom in the ℓ' cell. The quantity Φ_0 is the vibrational potential energy in the equilibrium state with all atoms located at their equilibrium positions and

$$\Phi_{\alpha\beta} = \left.\frac{\partial^2\Phi}{\partial u_\alpha\binom{\ell}{\kappa}\ \partial u_\beta\binom{\ell'}{\kappa'}}\right]_0 \quad . \tag{2.2}$$

The evaluation of these second derivatives is also to be made at the equilibrium state.

The equations of motion of the vibrating lattice can be found from the Hamiltonian (2.1a). In the small vibration harmonic approximation, the resulting equations are linear in the displacements u. Since boundary effects are generally uninteresting, one usually employs the Born-von Karman periodic boundary conditions in lattice vibration discussions. The periodic nature of the crystal implies that the αth component displacement of particle κ in all ℓ can be written as

$$u_\alpha\binom{\ell}{\kappa} = m_\kappa^{-1/2} u_\alpha(\kappa) \exp i[t\omega(k) + 2\pi\, k\cdot r(\ell)]. \quad (2.3)$$

When this is substituted into the equations of motion

$$m_\alpha \ddot{u}_\alpha\binom{\ell}{\kappa} = \sum_{\beta,\ell'} \sum_{\kappa'} \Phi_{\alpha\beta}\binom{\ell\ \ell'}{\kappa\ \kappa'} u_\beta\binom{\ell'}{\kappa'} , \quad (2.4)$$

one obtains a set of homogeneous equations whose solution exists only if the determinant of the coefficients of the displacements vanish. The matrix of the determinant is called the "dynamical matrix". The normal mode frequencies $\omega(k)$ associated with the wave vector k are solutions of

$$\det \{D_{\alpha\beta}\binom{k}{\kappa\ \kappa'} - \omega^2(k)\delta_{\alpha\beta}\delta_{\kappa\kappa'}\} = 0 \quad (2.5a)$$

where

$$D_{\alpha\beta}\binom{k}{\kappa\ \kappa'} = (m_\kappa m_{\kappa'})^{-1/2} \sum_\ell \Phi_{\alpha\beta}\binom{\ell}{\kappa\ \kappa'} \exp[2\pi i k\cdot r(\ell)] . \quad (2.5b)$$

The detailed dispersion relations

$$\omega = \omega(k) \quad , \quad (2.6)$$

which are the roots of the characteristic determinants (1.10a) as a function of k are very sensitive to the detailed choice of force constants Φ. Experiments such as the scattering of slow neutrons by crystals can be used to determine these dispersion curves[13]. The force constants $\Phi_{\alpha\beta}$ can sometimes be obtained from information on the elastic constants of the crystal. By combining both sets of data, a best set of force constants and dispersion curves can be found. Note that there are a number of branches of dispersion relations. For example, in the case of a monatomic cubic crystal, the dynamic matrix is 3x3 and there are three branches.

There is one model which leads to considerable simplification. It is a simple cubic lattice with nearest neighbor forces only,[6-8] both central and non-central, the non-central forces being required to keep the lattice stable, relative to shear. This case is simple because the x, y, and z components of the motion do not couple so that the dynamical matrix is diagonal. The equations of motion are those of a lattice with one degree of freedom per lattice point. We let the displacement of the $(\ell, m, n)^{th}$ particle from equilibrium in the x direction be

$$M\ddot{x}_{\ell,m,n} = \gamma_1(x_{\ell+1,m,n}-2x_{\ell,m,n}+x_{\ell-1,m,n})$$

$$+ \gamma_2(x_{\ell,m+1,n}-2x_{\ell,m,n}+ x_{\ell,m-1,n})$$

$$+ \gamma_3(x_{\ell,m,n+1}-2x_{\ell,m,n}+x_{\ell,n,n-1}) \qquad (2.7)$$

where γ_1 is the central force constant and γ_2 and γ_3 the non-central force constants between nearest neighbors. Two similar sets of equations exist for y and z displacements.

One can express the motions of particles in the lattice as linear combinations of the normal modes,

$$x_{\ell,m,n} = \exp\{i(\omega t+\ell\phi_1+m\phi_2+n\phi_3)\}, \qquad (2.8)$$

the ϕ's being chosen as

$$\phi_j = 2\pi k_j/N, \qquad k_j = 0, 1, 2, \ldots, N-1, \tag{2.9}$$

so that the x's satisfy periodic boundary conditions

$$x_{\ell,m,n} = x_{\ell+N,m,n}, \text{etc.}, \tag{2.10}$$

N being the number of lattice points in each direction in the lattice. The points (ϕ_1, ϕ_2, ϕ_3) are points on the reciprocal lattice. The normal mode frequencies are found to be the function $\omega(\phi_1, \phi_2, \phi_3)$ defined by

$$M\omega^2 = \sum_1^3 2\gamma_j(1 - \cos\phi_j), \tag{2.11}$$

which is triply periodic in reciprocal (ϕ_1, ϕ_2, ϕ_3) space. Equations similar to (2.7) exist for y and z components of the displacements as well. All yield the same dispersion relation. Each set of frequencies is called a branch of the frequency spectrum. When next nearest neighbor interactions are introduced, these branches become somewhat different. One frequency corresponds to each triple of ϕ's of the form (2.9).

The thermodynamic properties of a crystal depend on the normal mode frequencies $\{\omega\}$. For example, the specific heat at constant volume is

$$c_v = k \sum_j \left(\frac{h\omega_j}{2kT}\right)^2 / \sinh^2(\hbar\omega_j/2kT) \tag{2.12}$$

As $N \to \infty$ we see from (2.11) that the normal mode frequency becomes dense so that the sum (2.13) can be expressed as an integral over the frequency distribution function $g(\omega)$ which has the property $g(\omega)d\omega$ is the fraction of frequencies between ω and $\omega+d\omega$. Then

$$c_v = Nk \int_0^{\omega_L} g(\omega)\{(\hbar\omega/2kT)^2/\sinh^2(\hbar\omega/2kT)\}d\omega \tag{2.13}$$

where ω_L is the largest frequency, which, in case (2.11), becomes

$$M\omega^2 = 4(\gamma_1+\gamma_2+\gamma_3). \tag{2.14}$$

Every lattice point $(2\pi k_1/N, 2\pi k_2/N, 2\pi k_3/N)$ corresponds to a frequency ω. In the limit $N \to \infty$ the number of lattice points in the region $(0 < \phi_j < 2\pi)$, $j=1,2,3$ becomes dense and one can construct surfaces of constant frequency. In two dimensional cases, there are only two ϕ's, ϕ_1 and ϕ_2, and one has curves of constant frequency. These are exhibited in Fig. 4 as obtained from (2.11) with $\gamma_3 \equiv 0$, the dispersion relation obtained from the 2D equations of motion

$$\begin{aligned} M\ddot{x}_{\ell,m} &= \gamma_1(x_{\ell+1,m}-2x_{\ell,m}+x_{\ell-1,m}) \\ &+ \gamma_2(x_{\ell,m+1}-2x_{\ell,m}+x_{\ell,m-1}). \end{aligned} \tag{2.15}$$

If $G(\omega^2)$ is defined as the fraction of square frequencies between ω^2 and $\omega^2+d\omega^2$, then it is clear that $G(\omega^2)$ is proportional to the rate at which (ϕ_1,ϕ_2,ϕ_3) space is swept out by a surface of constant frequency as the frequency increases. Since

$$g(\omega)d\omega = 2\omega G(\omega^2)\, d\omega, \tag{2.16}$$

$g(\omega) = 2\omega G(\omega^2)$, and the distribution function $g(\omega)$ can be expressed (for a single branch of the spectrum) as the volume integral

$$g(\omega) = \frac{1}{8\pi^3}\frac{\partial}{\partial\omega}\int\int\int_{0<\omega(\phi_1\phi_2\phi_3)<\omega} d\phi_1\, d\phi_2\, d\phi_3 \tag{2.17a}$$

or the surface integral

$$g(\omega) = \frac{1}{(2\pi)^2}\int\int\frac{ds}{|\text{grad }\omega|} \tag{2.17b}$$

where the integration proceeds over the entire surface ω = constant.

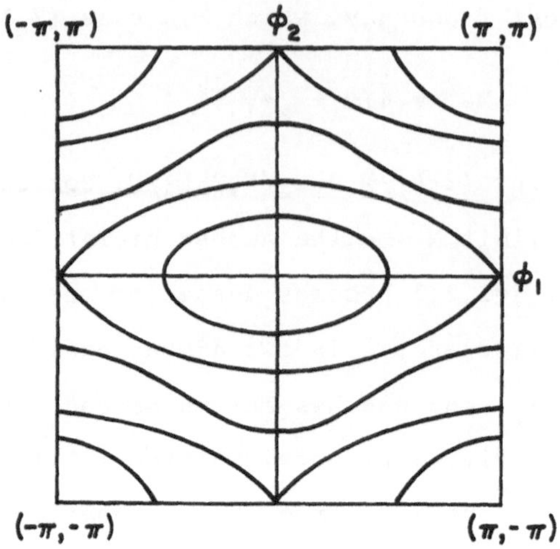

FIGURE 4. Schematic curves of constant frequency in the reciprocal lattice of a two dimensional crystal.

The frequency spectrum is easily obtained in the 2D case. One finds from the 2D analogue of (2.11) that

$$M^2\omega - 2(\gamma_1+\gamma_2) = -2\gamma_1\cos\phi_1 - 2\gamma_2\cos\phi_2 \text{ and } M\omega_L^2 = 4(\gamma_1+\gamma_2). \quad (2.18)$$

The lines of constant frequency are plotted in fig. 4. If, when $\omega^2 < \frac{1}{2}\omega_L^2$, we multiply by 4 the fraction of frequencies in the first quadrant whose square is between ω^2 and $\omega^2 + d\omega^2$, we find

$$G(\omega^2) = \frac{4}{4\pi^2}\frac{\partial}{\partial\omega^2}\int_0^{\phi} d\phi_1 \int_0^{\phi_2} d\phi_2 \quad (2.19)$$

where ϕ_2 is to be expressed in terms of ϕ_1 through (2.18) and ϕ is the value of ϕ_1 when $\phi_2 = 0$, i.e.,

$$2\gamma_1\cos\phi = 2\gamma_1 - M\omega^2$$

or

$$\phi = \cos^{-1}\left(1 - \frac{M\omega^2}{2\gamma_1}\right)$$

Then

$$G(\omega^2) = \frac{1}{\pi^2} \int_0^{\phi} d\phi_1 \left\{1 - \left[\frac{2(\gamma_1+\gamma_2)-M\omega^2-2\gamma_1\cos\phi_1}{2\gamma_2}\right]^2\right\}^{-1/2}$$

so that if we define a new variable of integration, x, by

$$(x-1)(M\omega^2/4\gamma_1) = \cos\phi_1 - 1$$

we see that

$$G(\omega^2) = \frac{1}{(\pi\omega)^2} \int_{-1}^{1} \left\{\left(1-x^2\right)\left[\left(\frac{8\gamma_1}{M\omega^2} - 1\right) + x\right]\left[\left(\frac{8\gamma_2}{M\omega^2} -2\right) - x\right]\right\}^{-1/2} \quad (2.20a)$$

which is a complete elliptic integral of the second kind. This is defined by

$$K(k) = \int_0^{\pi/2} (1 - k^2 \sin^2\theta)^{-1/2} d\theta \, . \quad (2.20b)$$

One finds that

$$G(\omega^2) = \frac{M}{2\pi^2(\gamma_1\gamma_2)^{1 2}} K\left(\frac{M\omega(\omega_L^2-\omega^2)^{1/2}}{4(\gamma_1\gamma_2)^{1/2}}\right) \text{ if } 0 < M^2\omega^2(\omega_L^2-\omega^2) < 16\gamma_1\gamma_2 \, . \quad (2.21a)$$

In a similar manner it can be shown that

$$G(\omega^2) = \frac{2}{\omega\pi^2(\omega_L^2-\omega^2)^{1/2}} K\left(\frac{2(\gamma_1\gamma_2)^{1/2}}{\omega(\omega_L^2-\omega^2)^{1/2}}\right) \text{ if } \omega^2(\omega_L^2-\omega^2) > 16\gamma_1\gamma_2 \quad (2.21b)$$

where throughout $M\omega_L^2 = 4(\gamma_1+\gamma_2)$. There are two logarithmic singularities in $G(\omega^2)$, one at $M\omega^2 = 4\gamma_1$ and the other at $M\omega^2 = 4\gamma_2$. The functions $G(\omega^2)$ and $g(\omega)$ are plotted in Fig. 5.

While one cannot obtain simple formulae such as (2.21) for $G(\omega^2)$ in the 3D case, a representation as a single integral can be found and $G(\omega^2)$ can be easily calculated numerically. The graph of the function $g(\omega)$ for the simple cubic lattice model discussed above is given in Fig. 6. Fig. 7 is that for a more realistic model of sodium which is constructed to fit data on neutron scattering.

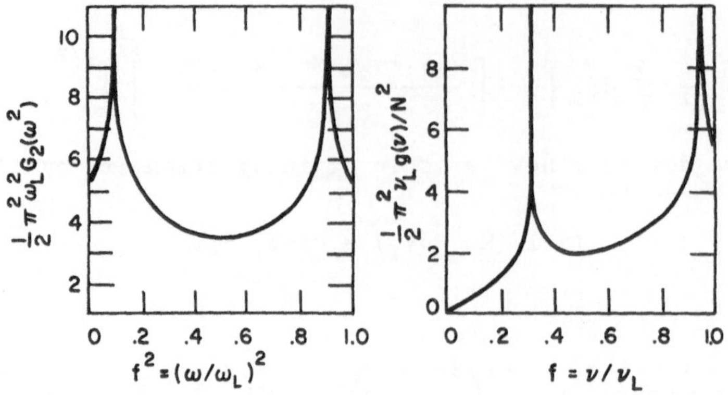

FIGURE 5. Frequency spectrum of a 2-D lattice with $\gamma_1/\gamma_2 = 1/9$. Logarithmic singularities occur at $f = 0.316$ and 0.948.

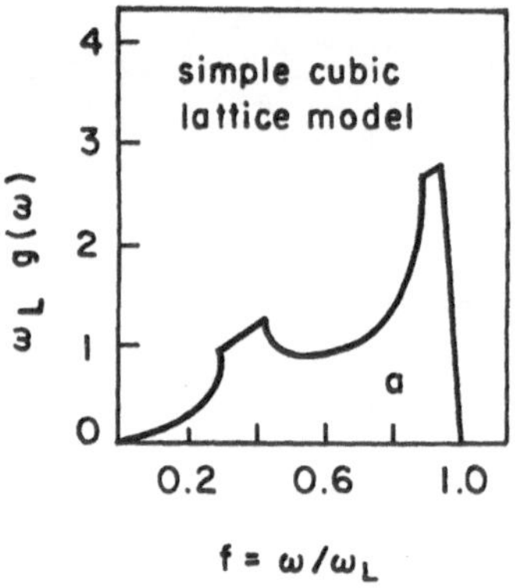

FIGURE 6. (a) Frequency spectrum of a simple cubic model lattice with nearest neighbor central and non-central forces. The central forces are 8 times as large as non-central ones.

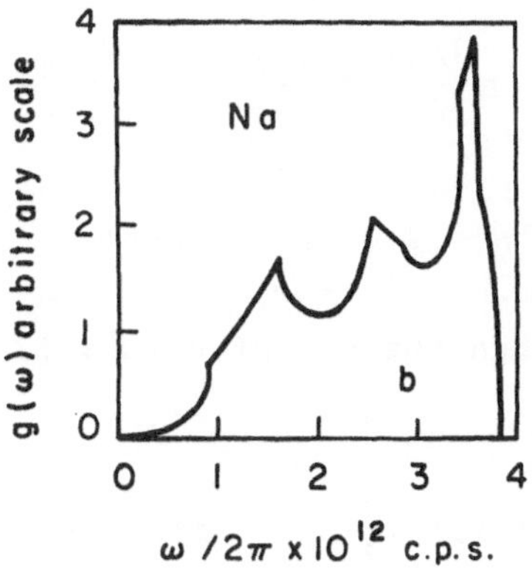

FIGURE 7. (b) Experimental frequency spectrum of sodium

The singularities are of the form $(|\omega-\omega_c|)^{\frac{1}{2}}$ as ω approaches certain critical values ω_c. A considerable literature exists on the nature of these singularities and the reason for their existence.[5,6,9-12]

A considerable literature also exists on the experimental determination of dispersion curves for various crystals[13].

The frequency spectrum and dispersion curves for polyatomic crystals is somewhat more complicated than that for monatomic ones.[1,5] In a simple cubic lattice in which the two atomic species alternate along the lattice points (for example, an NaCl type lattice), a generalization of (2.11) can be found for the squares of normal mode frequencies[14]. If the light and heavy masses are, respectively, M_1 and M_2, there are two branches to the frequency spectrum. The high frequency branch, which is known as the optical branch, is

$$\omega_+^2 = \tfrac{1}{2}(\omega_1^2+\omega_2^2) + \tfrac{1}{2}[(\omega_2^2-\omega_1^2)^2 + 16X^2/M_1M_2]^{\frac{1}{2}} \qquad (2.22a)$$

and the low frequency branch, which is known as the acoustical branch, is

$$\omega_-^2 = \tfrac{1}{2}(\omega_1^2+\omega_2^2) - \tfrac{1}{2}[(\omega_2^2-\omega_1^2)^2 + 16X^2/M_1M_2]^{1/2} \quad (2.22b)$$

where

$$X = \sum_{j=1}^{3} \gamma_1 \cos\phi_j \quad (2.23)$$

and the ϕ_j's run through the set of values (2.9).

Also

$$M_j\omega_j^2 = 2(\gamma_1 + \gamma_2 + \gamma_3) \qquad \text{with } j = 1,2 \quad (2.24)$$

The frequency distribution in this case is plotted in Fig. 8. Note the band gap between the two bands.

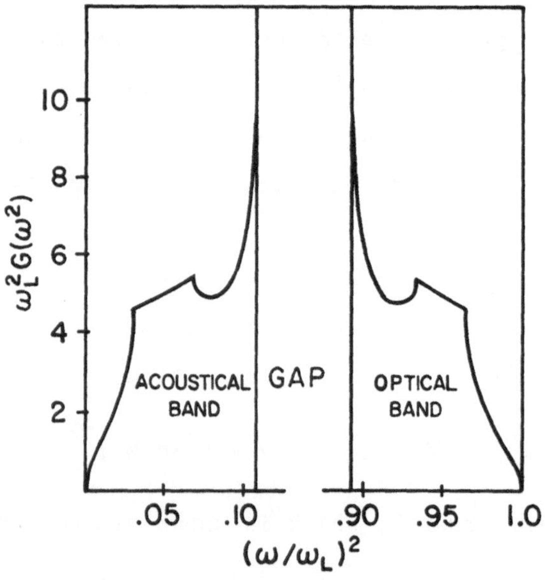

FIGURE 8. Distribution of ω^2 for diatomic lattice with $M_2 = 8M_1$ and $\gamma_1/\gamma_2 = 8$.

CHAPTER III. PROPAGATION OF A PULSE IN A TWO DIMENSIONAL ANHARMONIC LATTICE.

Considerable insight on the propagation of a disturbance in two- and three-dimensional lattices can be obtained from machine calculations recently made by Payton, Rich, and Visscher[15] on 2D square lattices with linear and non-linear force laws in the special case in which all interactions are between nearest neighbors only (with both central and non-central forces). The linear case is the model which we have discussed above. In the non-linear case, the force law chosen was an expansion to fourth order of the Lennard-Jones potential

$$\begin{aligned} \Phi_0(r) &= \varepsilon_0[(r/r_0)^{12} - 2(r_0/r)^6] \\ &\simeq -\varepsilon_0 + \tfrac{1}{2}\gamma(r-r_0)^2 - \tfrac{1}{3}\mu(r-r_0)^3 + \tfrac{1}{4}\nu(r-r_0)^4 . \end{aligned} \tag{3.1}$$

The magnitudes of μ and ν relative to γ were taken to be appropriate for noble gas solids in the anharmonic case, and zero in the harmonic one. With arbitrary units such that $\gamma=1$, the potentials used were

$$\Phi(x) = \tfrac{1}{2}x^2 - \tfrac{1}{3}\mu x^3 + \tfrac{1}{4}\nu x^4 . \tag{3.2}$$

Periodic boundary conditions were chosen in the direction normal to that of the propagation of the input disturbance and reflecting boundary conditions were set at the end of the lattice in the direction of propagation of the disturbance so that it could be reflected from the ends.

The feature that has made the calculations especially interesting is that they have been exhibited on a movie film. The 2D lattice is represented by a grid and the energy at a lattice point is shown by the raising of the grid at that lattice point by an amount proportional to the energy (c.f. Fig. 9). The initial disturbance was the same at all lattice points along lines normal to the direction of propagation. The initial energy pulse had a kinetic energy per atom corresponding to 3 K and extended over several lattice rows.

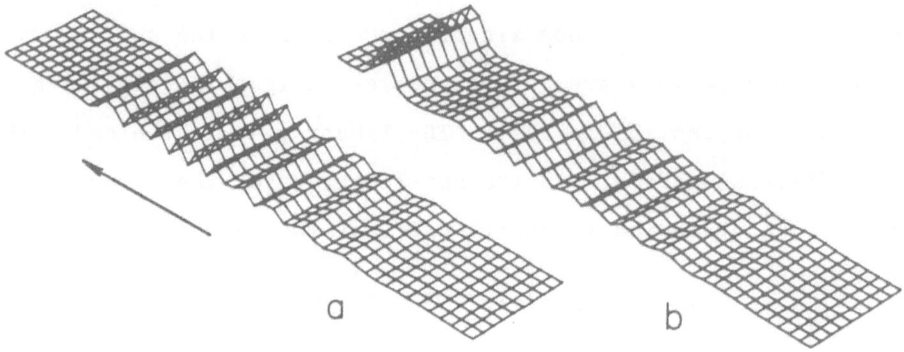

FIGURE 9. Comparison of the energy waves at equal times in harmonic (a) and anharmonic (b) monatomic lattices.

Two typical film frames are shown in Figs. 9a and 9b; 9a represents a stage in the propagation of an energy pulse through a monatomic harmonic lattice. The initial pulse diminishes in amplitude as it progresses and, due to the dispersive character of the lattice, a trail of smaller amplitude waves develops behind it. Figure 9b represents the situation as the wave propagates through the anharmonic lattice postulated above. Figures 9a and 9b correspond to the states of development at the same time on the two lattices. Notice that a shock wave with a sharp front appears in the anharmonic case. As expected, its propagation velocity exceeds that of the small amplitude wave which trails it. The small amplitude trailing part behaves in essentially the way that a wave sould propagate in the harmonic lattice. Payton, Rich, and Visscher have also made movies of the propagation of waves in lattices with defects. These are discussed in Chapter IV.

Similar calculations can also be made in 3D lattices but a film presentation is not possible.

A fundamental difficulty arises in the development of a theory of the propagation of large amplitude or shock waves in a 3D lattice. When large amplitude displacements from equilibrium occur, atoms exchange positions and lattice imperfections appear. These are very hard to program into a calculation.

A strong shock in a real solid has a front of only a few atomic layers. The mechanism of the rearrangement of the atoms from a lower to a higher density state is unclear as is the dissipation mechanism after the shock passes. Cyril Smith has proposed that a shock front contains an array of edge dislocations which move with the shock front and which account for the increase in density in the shock without the destruction of the lattice.

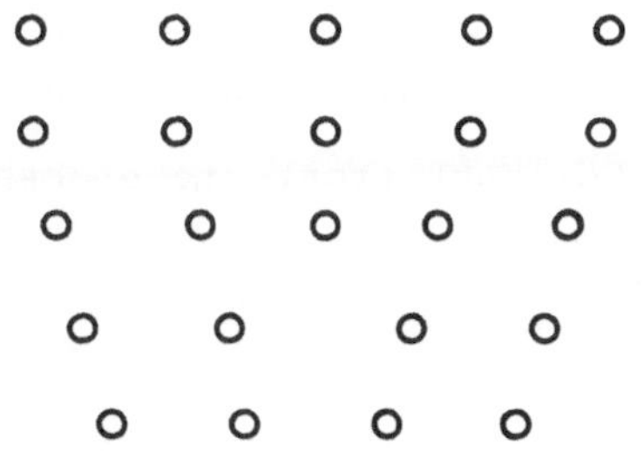

FIGURE 10. Edge dislocation. Symbol ⊥

A schematic picture of an edge dislocation is shown in Fig. 10 and the type of array of edge dislocation which might appear in the shock front is given in Fig. 11.

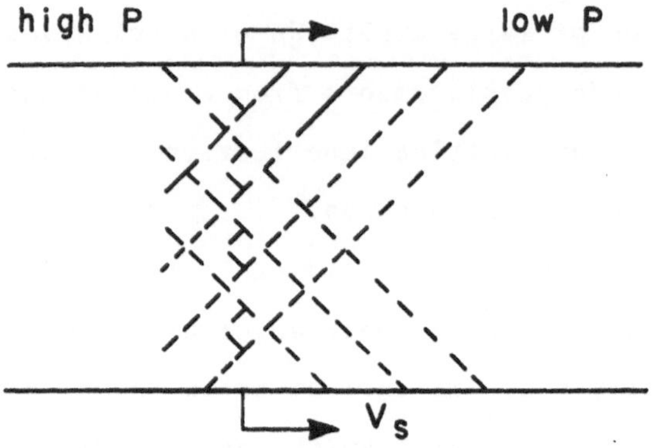

FIGURE 11. Development of dislocation lines at shock front. Region 2, left of shock front, is high pressure. Region to the right is low pressure.[4]

Gilman and Vineyard[4] have discussed this model to some extent. The molecular motions in the shock front are probably similar to those experienced in melting.

CHAPTER IV. EFFECT OF DEFECTS ON LATTICE VIBRATIONS

Let us now investigate the effect of defects on lattice vibrations and on the manner in which waves propagate through a defective lattice. Impurities are an important class of defects as are dislocations. The replacement of an atom in a lattice by an impurity corresponds to a local variation in mass and force constants, while a dislocation is essentially an extended variation in force constants.

Even without introducing a detailed model for the impurity, several effects can be deduced on general grounds by applying certain theorems (first due to Rayleigh) concerning systems of springs and masses. Let us suppose that the normal mode frequencies of an unperturbed lattice are

$$\omega_1^{(0)} < \omega_2^{(0)} < \omega_3^{(0)} < \omega_4^{(0)} < \ldots < \omega_n^{(0)} . \qquad (4.1)$$

Then, if one mass in the system is reduced, all frequencies are increased; however, the jth, $\omega_j^{(1)}$, is bounded between the unperturbed j^{th} and $(j+1)^{st}$ so that

$$\omega_j^{(0)} < \omega_j^{(1)} < \omega_{j+1}^{(0)} \qquad j = 1,2,3\ldots,n-1 \qquad (4.2a)$$

$$\omega_n^{(0)} < \omega_n^{(1)} \qquad (4.2b)$$

If, instead, one mass is increased, one obtains the new set of frequencies $\omega_j^{(2)}$ such that

$$\omega_j^{(0)} > \omega_j^{(2)} > \omega_{j-1}^{(0)} \qquad j = 1,2,\ldots,n-1 \qquad (4.3a)$$

$$\omega_1^{(0)} > \omega_1^{(2)} . \qquad (4.3b)$$

The increase in a force constant has the same effect as the decrease of a mass (and vice versa). These results are, of course, in qualitative agreement with those involving a mass tied to a rigid wall by a spring in that a decrease in mass increases the frequency of vibration as does an increase in spring constant, and vice versa.

As we observed in the last section, a crystal contains a large number of degrees of freedom and, therefore, a large number of normal mode frequencies. These were shown to appear in dense bands. The inequalities (4.2) and (4.3) then imply that the frequencies of the perturbed lattice are essentially the same as those of the unperturbed one except that frequencies at the band edge might be displaced a considerable distance from the band. For example, if the lattice contains a light defect, the frequency $\omega_n^{(1)}$ (see inequality 4.2b) is not bounded from above so that it might become separated from the band. Indeed, if a linear chain which has no imperfections has a highest frequency ω_L, it can be shown that the defect frequency which escapes from the band and which is due to the light mass defect, is[16]

$$\omega = \omega_L [Q(2-Q)]^{-1/2} \qquad \text{with} \qquad Q = m'/m, \qquad (4.4)$$

m' being the light defect mass and m the mass of a host lattice atom.

An observation first made by Lord Kelvin is useful for the interpretation of the nature of the normal mode of vibration associated with a frequency that is separated from the band. He found that if one tries to drive a wave into a periodic structure from one end with a frequency that is not in the frequency band, the wave damps out in a distance that depends on the displacement of the driven frequencies from the band edge, the penetration depth decreasing as the displacement from the band edge increases.

At a given temperature, thermal motion drives the various normal modes. One whose frequency lies out of the band remains localized because from Kelvin's theorem it cannot propagate far in the crystal. Hence in a monatomic crystal, localized modes develop around defects which involve light masses or force constants larger than those associated with pairs of atoms in an unperturbed crystal.

If two mass defects are far from each other, a localized mode develops around each and both modes have the same frequency. However, as the two are brought closer to each other until the ranges of localization overlap, the two modes interact with each other and the frequency degeneracy is split, one frequency going up and the other down. An impurity generally corresponds to a change in mass and several force constants; six in the case of a simple cubic lattice, with nearest neighbor interaction only. Since the splitting of the degeneracy by variation of several force constants is analogous to that by change of several masses, under favorable conditions one local mode might appear for the mass change and six with similar frequencies for the force constant change in a simple cubic lattice. As the concentration of impurities increases, impurity bands of frequencies develop.

As was mentioned in Section II, in the case of ordered diatomic and polyatomic lattices, the frequency spectrum contains optical (high frequency) and acoustical (low frequency) bands. A typical defect in such lattices is an interchange of two atoms, which corresponds to a local disorder. The change in frequency spectrum due to interchanging A and B atoms in an ordered AB lattice can be seen from a consideration of Rayleigh's theorem[14]. Let the mass of an A atom be M_A and that of the B atom be M_B with $M_A < M_B$. Then the act of replacing a heavy B atom by a lighter A atom causes a localized mode to emerge from the top of the optical and another from the top of the acoustical band. The completion of the interchange by replacing the light A with a heavier B causes a mode to emerge from the bottom of the optical band. Similar remarks can be made about the influence of a change in force constants. The various frequencies of the localized modes and their sources are sketched in Fig. 12 for the case of $\gamma_{AA} > \gamma_{AB}$ and $\gamma_{BB} < \gamma_{AB}$. The diagram corresponds to our simple cubic lattice model which possesses one degree of freedom per

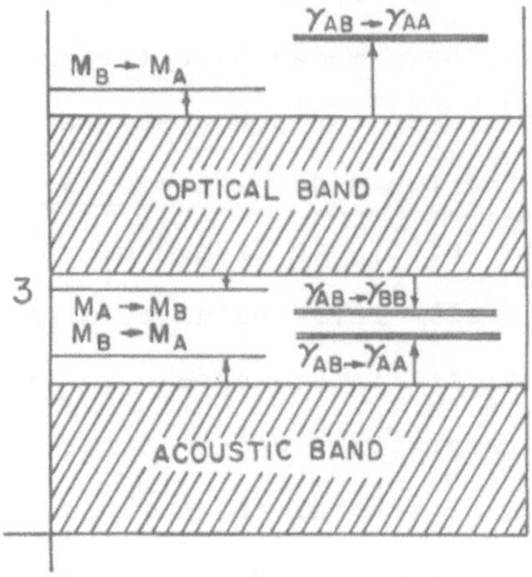

FIGURE 12. Sketch of localized mode frequencies in a diatomic lattice which result from a disordered pair of atoms $M_B \to M_A$ is used to identify the mode that results from replacing the mass of an atom of kind B by that of one of kind A. $M_A < M_B$.

lattice point. All degrees of freedom are taken into account by multiplying each mode by a degeneracy factor of three. This degeneracy is split by considering next nearest neighbor interactions. If two force constants are changed at widely separated points in a lattice, the associated localized mode frequencies are degenerate. The degeneracy is split as the two anomalous force constants are brought closer together. If it is assumed that only central force constants are changed when an A atom is replaced by a B atom in our ordered diatomic lattice, two new force constants are associated with the substituted B atom. The pairs of closely lying frequencies in Fig. 12 are drawn to correspond to the resulting splitting of pairs of anomalous force constant localized mode frequencies.

Some of the localized modes shown in Fig. 12 might be suppressed for the following reason. Suppose a small decrease in a mass is made so that frequencies barely rise from the bands. A large decrease in the force constants which reduce all frequencies might return these modes to the bands. This interplay between changes of masses and force constants has been discussed for one-dimensional systems elsewhere. Situations exist in which a frequency does not emerge from the bands until a parameter is changed by more than a certain critical amount.

Now consider a two-or three-dimensional lattice with a low concentration of randomly distributed mass defects. Let a disturbance propagate through the lattice and suppose that the disturbance extends over the complete line, (or plane, in the 3D case) normal to the direction of porpagation while the depth of the disturbance is of the order of five to twenty-five lattice spacings. One would expect the following events to occur.

In the neighborhood of a light impurity, a localized mode would be generated. However, since the frequency of vibration of the light particle is higher than that of its neighbors, they would tend to be in phase with the pulse for a long time while the light impurity would be sometimes in phase and sometimes out of phase with the pulse. Hence the light impurity would not couple to and pick up energy from the pulse as well as its heavier neighbors would. If, for example, one plotted the energy associated with each particle, the light mass would have less than its neighbors. On the other hand, a heavy defect has greater inertia than its neighbors so that it remains in phase with the driving pulse for a longer time and can pick up more energy leading to a spike in an energy curve. The motion of the heavy atom is called a resonance mode. It has a finite lifetime which is the time required to transmit its energy to the rest of the crystal. Changes in force constants yield similar results--an increase in force constant corresponding to a decrease in mass, and vice versa.

Defects act as scatterers so that part of the pulse is reflected backward by them and, indeed, at a fixed concentration in a sufficiently long sample, one would expect little of the pulse to continue through it without being reflected back.

The mathematical theory of the effect of defects has been developed in a number of papers, including those in References 16 and 17. Several reviews exist (see, for example, References 5, 18 and 19). Recent experimental work is reported in Reference 20.

Payton, Rich, and Visscher[15], in their film program discussed in the last section, have given an excellent visual presentation of the propagation of a disturbance in a lattice with defects. The atoms of the host lattice were given a mass three, and light and heavy defects were given masses one and nine, respectively.

The first cases considered were an isolated light and an isolated heavy defect in an otherwise perfect two-dimensional lattice. The quantity exhibited in Figure 13 which summarized their results is the energy at each lattice point.

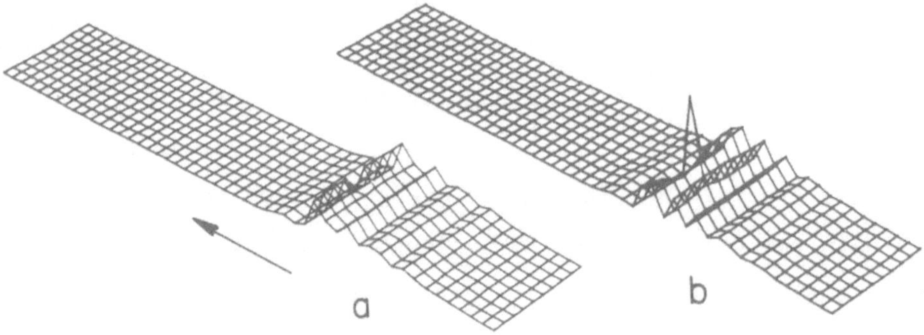

FIGURE 13. Illustrations of the dip occurring in the energy wave as it passes over a light impurity (a) and the spike resulting from encounter with a heavy impurity (b).

The difference between the passage of a wave through a light and a heavy impurity is shown in Fig. 13. The dip in energy in the light impurity case and the spike in the heavy one are as described above.

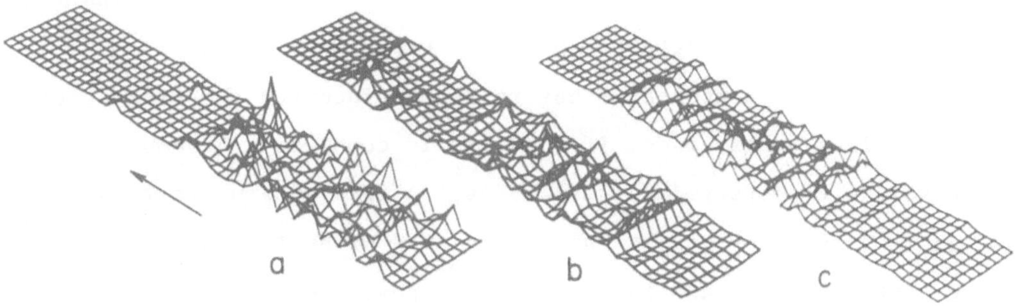

FIGURE 14. Equal time comparison of energy penetration into a harmonic lattice containing 15% heavy impurities (a) with that into a 15% light impurity region (b). Impurity sites are identical in the two pictures. (c) corresponds to heavy impurities in anharmonic cases.

Figures 14a and 14b give an equal time comparison of energy penetration into a harmonic lattice containing 15% heavy atoms with that into a 15% light impurity region. The impurity sites are the same in both cases. The pulse seems to propagate further in the lattice with light impurities than in the one with heavy. Apparently the heavy impurities reflect the incident wave better than do the light ones.

Finally, Figs. 14a and 14c show the effect of 15% heavy impurities on both harmonic and anharmonic lattices. Clearly the pulse propagates through defects in the anharmonic lattice (case c) more easily than through the harmonic one.

All the figures were taken from reference 15. The movies exist in the Los Alamos film library.

CHAPTER V. ERGODICITY AND PERMANENT WAVES IN ANHARMONIC ONE-DIMENSIONAL CHAINS

One of the basic ideas in the classical statistical mechanics of systems undergoing small vibrations is that, at equilibrium at a given temperature, the energy of the system becomes equally divided into the various normal modes of vibration. This is the so-called equipartition theorem which states that the energy in every mode is ($\frac{1}{2}kT$).

Of course some mechanism has to be provided for the weak coupling of the various modes since, if a system is completely harmonic, energy can never be transferred between modes. A weak anharmonicity, a radiation field, or contact with some kind of heat bath are considered to be sufficient to make the mode mixing possible.

The equipartition theorem is deduced from equilibrium statistical mechanics and not from an investigation of the asymptotic behavior of its dynamics. Hence one could imagine (but not many did seriously) that equilibrium might never be achieved, in which case the theorem would not be applicable. A number of calculations have been made recently, investigating this point.

A. Fermi, Pasta, and Ulam Calculations.[21]

In the early days of high speed computers, E. Fermi became interested in their employment for the solution of non-linear problems. He felt that future fundamental theories in physics may involve non-linear operations and equations and that it would be useful to develop some experience in this field. As a test problem, he thought that the dynamics of the approach to equipartition would be interesting to investigate.

A problem which Fermi, Pasta, and Ulam investigated with MANIAC I at Los Alamos was the ergodic behavior of a linear chain of particles which interacted through a non-linear interparticle force. The interaction laws studied were respectively quadratic, cubic, and certain broken line interactions so that the equations of motion studied successively were

$$\ddot{x}_i = (x_{i+1} + x_{i-1} - 2x_i) + \alpha[(x_{i+1} - x_i)^2 - (x_i - x_{i-1})^2] \qquad (5.1a)$$

$$\ddot{x}_i = (x_{i+1} + x_{i-1} - 2x_i) + \beta[(x_{i+1} - x_i)^3 - (x_i - x_{i-1})^3] \qquad (5.1b)$$

$$\ddot{x}_i = \delta_1(x_{i+1} - x_i) - \delta_2(x_i - x_{i-1}) + c \qquad i = 1, 2, 3 \ldots N \quad (5.1c)$$

where x_i represents the displacement of the ith atom from its equilibrium position. The constants α and β were chosen so that at maximum displacement the non-linear terms were only about one-tenth of the linear ones. In the third case, the parameters δ_1, δ_2, and c were not constants but assumed different values depending upon whether or not the quantities in the parentheses were greater or less than a certain value fixed in advance. The values of N used by FPU were 16, 32, and 64.

The total energy of the chain, in the harmonic approximation, is

$$E = \tfrac{1}{2}\sum_j \{\dot{x}_j^2 + [(x_{j+1} - x_j)^2 + (x_j - x_{j-1})^2]\}$$

$$= \sum_k \{\tfrac{1}{2}\dot{a}_k^2 + 2a_k^2 \sin^2(\pi k/2N)\} \qquad (5.2)$$

The normal coordinates a_k are defined by

$$\ddot{a}_k(t) = \sum_j x_j(t)\sin(kj\pi/N) \qquad (5.3)$$

so that

$$x_j(t) = \frac{2}{N}\sum_{k=1}^{N} a_k(t)\sin(jk\pi/N) \qquad (5.4)$$

Also

$$a_k + \omega_k^2 a_k = 0 \qquad k = 1,\ldots,N \qquad (5.5a)$$

where

$$\omega_k = 2\sin(\pi k/2N) \qquad (5.5b)$$

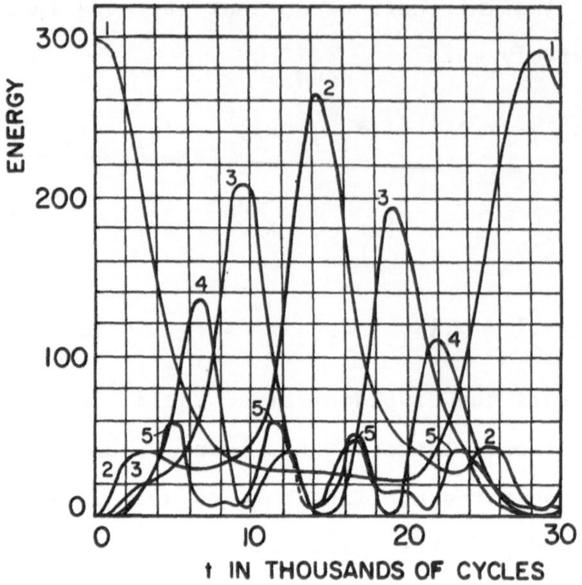

FIGURE 15. Variation in energy in various modes as a function of time. The units of energy are arbitrary. $N = 32$; $\alpha = 1/4$. The initial form of the chain was a single sine wave. Modes higher than the 5th never exceeded 20 units of energy on the scale given here.[21]

When anharmonic terms are included their contribution, which in case (5.1a) is proportional to the sum of $\alpha(x_{j+1} - x_i)^3$ and in case (5.1b) to sum of $\beta(x_{j+1} - x_j)^4$, must be added to (5.2). In the FPU calculation, their contribution to E was never more than a few percent. In the presence of anharmonicities, the various a_k's couple so that the members of the set of equations (5.5) are all coupled.

If all the energy is initially in the normal mode $k = 1$, one would expect the coupling to generate a slow flow of energy into the higher modes until equipartition accompanied by some small fluctuation is achieved. As $N \to \infty$ these fluctuations should vanish.

In the FPU calculations, the process started as expected, energy flowed into the second, then the third, then the fourth and fifth mode, but, to the surprise of all concerned, most of the energy suddenly flowed back into the second mode and soon into the first mode with this exchange continuing. With $N = 32$ and $\alpha = \frac{1}{4}$, the total energy in modes with $k > 5$ never exceeded 8%. The details are shown in Fig. 15.

B. Some Aspects of Perturbation Theory of Chain with Quadratic Non-linearity.

While perturbation theory is not the most effective way of obtaining a clear understanding of non-linear processes, one can without too much difficulty obtain some ideas from it. The easiest case to discuss is the quadratic non-linearity which yields the equations of motion

$$m\ddot{x}_j = \gamma(x_{j-1} - 2x_j + x_{j+1}) + \alpha[(x_{j-1} - x_j)^2 - (x_j - x_{j+1})^2] \quad (5.6)$$

Let us consider a ring of springs and masses so that $x_{j+N} \equiv x_j$. Then we set

$$x_j = \frac{1}{N} \sum_{k=1}^{N} a_k \exp(2\pi ikj)/N \quad (5.7a)$$

so that

$$(x_{j-1} - x_j) = \frac{1}{N} \Sigma\, a_k[-1 + \exp(-2\pi ik/N)] \exp(2\pi ikj/N) \quad (5.7b)$$

and

$$(x_{j-1} - x_j)^2 = \frac{1}{N^2} \sum_k \sum_{k'} a_{k'} a_{k-k'} e^{2\pi ijk/N} [1 - e^{-2\pi ik'/N}][1 - e^{-2\pi i(k-k')/N}] \quad (5.7c)$$

Similar expressions exist for $(x_j - x_{j+1})$ and $(x_j - x_{j+1})^2$. If all these equations are substituted into (5.6) and coefficients of $\exp(2\pi ikj/N)$ on both sides of the resulting equation are equated, one finds that

$$\ddot{a}_k + \omega_k^2 a_k = (8i\alpha/mN) \sum_{k'} a_{k'} a_{k-k'} \sin(\pi k/N) \sin(\pi k'/N) \sin\pi(k'-k)/N \quad (5.8a)$$

where

$$\omega_k^2 = (2\gamma/m)(1 - \cos 2\pi k/N) \tag{5.8b}$$

or

$$\omega_k = \omega_L \sin\pi k/N \quad \text{with} \quad \omega_L^2 = 4\gamma/m. \tag{5.8c}$$

Since the solution of

$$\ddot{a}_k + \omega_k^2 a_k = f(t) \tag{5.9a}$$

is

$$a_k(t) = a_k(0)\cos t\omega_k + \omega_k^{-1}\dot{a}_k(0)\sin t\omega_k + \omega_k^{-1}\int_0^t f(\tau)\sin(t-\tau)\omega_k \, d\tau, \tag{5.9b}$$

we see that the differential equation (5.9a) is equivalent to the non-linear integral equation

$$a_k(t) = a_k(0)\cos t\omega_k + \omega_k^{-1}\dot{a}_k(0)\sin t\omega_k$$
$$+ (8i\alpha/mN\omega_L)\int_0^t \sin(t-\tau)\omega_k \sum_{k'=1}^{N} a_{k'}(\tau)a_{k-k'}(\tau)\sin\frac{\pi k'}{N}\sin\frac{(k-k')}{N} \, d\tau . \tag{5.10}$$

A systematic but tedious way of solving this equation is to iterate to obtain a power series in the small parameter α. To get some idea of how the mode coupling develops, let us choose the simple example $\dot{a}_k(0) = 0$ for all k and

$$a_k(0) = Nc(\delta_{k,1} - \delta_{k,N-1})/2i \quad . \tag{5.11a}$$

This corresponds to the initial sine distribution

$$x_j(0) = c \sin(2\pi j/N) \quad . \tag{5.11b}$$

Note that

$$\omega_k = \omega_{N-k} \text{ and } \omega_k = -\omega_{N+k}, \ \omega_k = -\omega_{-k} \; . \tag{5.12}$$

Then upon iteration we find that

$$a_k(t) = \frac{Nc}{2i}(\delta_{k,1} - \delta_{k,N-1}) \cos t\omega_1$$

$$-\frac{2i\alpha c^2 N\omega_1^2}{m\omega_L^3} \sin^2\frac{\pi}{N} (\delta_{k,2} - 2\delta_{k,N} + \delta_{k,N-2}) \int_0^t \cos^2 \tau\omega_1 \sin(t-\tau)\omega_k \, d\tau$$

$$+ 0(\alpha^2 c^4) \qquad (5.13)$$

The integral is elementary and one finds

$$a_k(t)/N = (c/2i)(\delta_{k,1} - \delta_{k,N-1}) \cos t\omega_1$$

$$- [(i\alpha/4\gamma)c^2\omega_1^2/\omega_L\omega_k](\delta_{k,2} + \delta_{k,N-2})\{(1-\cos t\omega_k)$$

$$+ [\omega_k^2/(\omega_k^2 - 4\omega_1^2)](\cos 2t\omega_1 - \cos t\omega_k)\} + 0(\alpha^2 c^4). \qquad (5.14)$$

We have used the fact that $\omega_1^2 = \omega_L^2 \sin^2\pi/N$, and the term proportional to $\delta_{k,N}$ has been dropped because the quantity in the bracket vanishes when $k = N$ (since $\omega_N = 0$). The only nonvanishing a_k's are a_1, a_{N-1}, a_2 and a_{N-2}. Hence, the first order perturbation only excites the second sine mode, $\sin 4\pi j/N$. If one iterates again, the terms of order $\alpha^2 c^4$ correspond to excitation of the third sine mode, etc. Higher modes appear with coefficients that are higher powers of $\alpha^2 c^4$.

In the FPU calculation, c was chosen to be of order 1 and α of order 1/10. Hence it would be hard for the higher modes to becomes excited. On the other hand, they could get excited through resonances. Note the frequency denominator $\omega_k^2 - 4\omega_1^2$. As one develops perturbation theory to higher and higher order, denominators of the form $n_k^2\omega_k^2 - n_\ell^2\omega_\ell^2$ appear where n_k and n_ℓ are integers. Hence, if frequencies are commensurable so that $n_k\omega_k = n_\ell\omega_\ell$, the resonances appear and energy is easily transferred from the kth to the ℓth mode. Since as $\mu \to \nu$ $(\mu-\nu)^{-1}(\cos t\nu - \cos t\mu) \to t \sin t\mu$. This means that when a resonance exists, by waiting long enough the factor t eventually overwhelms the smallness of the factor $(\alpha c^2)^2$. This observation has been made by Ford[22,23] who also noticed that when N is a prime or a power of 2, no resonances exist in the normal mode frequencies. Since N was chosen to be 16, 32, or 64 in the FPU calculations, this efficient energy transfer mechanism did not exist and the energy spilled back from the second, third, or fourth modes into lower ones before higher ones ever had a chance to become excited.

Incidentally, resonance phenomena have been known for many years in celestial mechanics. Newton's theory of gravitation and theory of planetary orbits was under attack for many years by astronomers who noted that the orbit parameters of Saturn and Jupiter seemed to vary linearly with the time. The enigma of the "mean motion" of these planets was resolved by Laplace who observed that the small value of $5\omega-2\omega'$ (ω and ω' being the unperturbed frequencies of the orbits of Saturn and Jupiter) led to a resonance. The period of the coupled system was 929 years. An interesting discussion of resonances (especially between asteroids) in the solar system was given by E. W. Brown[24].

The solar system also teaches us that one should not be too surprised that equipartition does not occur in all systems of slightly anharmonic oscillators. Each unperturbed planetary orbit can be considered to be a normal mode of the system. The weak gravitational forces between planets furnish the weak nonlinear coupling. Since there seems to be no evidence of the equipartition of energy among the planets, more is required for the equipartition of energy in a closed system of oscillators than a slight anharmonic coupling between them.

C. Calculations of Northcote and Potts[25].

In order to examine the importance of the number of particles in the chain and to check the sensitivity of the FPU results to the nature of the model, Northcote and Potts investigated the model of a line of rigid spheres of diameter D connected by simple harmonic springs. The non-linearity is apparent only as an infinite repulsion when the spheres are in contact. This is an easy model to program for a computer because, between collisions, the solution of the equations of motion is known. One would start with an initial set of positions and momenta of the rigid spheres and, from the known solution of the harmonic problem, determine the new positions and momenta at some time t_1. If these positions and momenta indicate that no collisions occurred in this time t_1, a new set is found appropriate for a time t_2. Suppose it is clear that a collision between the ℓth and $\ell+1^{th}$ spheres has occurred in the interval t_2-t_1. Then one chooses a new time $t_1 < t_2' < t_2$ and determines whether the collision occurs in the interval $t_2'-t_1$ or t_2-t_2'. This process can be continued until the collision time is determined to within any desired accuracy. In terms of the moment of collision, a new set of solutions of the equation of motion is developed with initial conditions obtained by interchanging the momenta of the ℓth and $\ell+1$'s particles and giving the other variables the values they had at the moment of collision.

The numerical results were quite different from those obtained by FPU (the same end conditions, particles at ends kept fixed, were used). Equipartition was achieved slowly when the chain started in the lowest mode and more rapidly when it started in a higher mode. The mixing of modes seemed to start effectively at the chain ends. The first collision of the atoms next to the ends with the end atoms gave a strong localized reflection so that the chain configuration after the collision required higher components of the harmonic normal modes for their description. The mixing does not have to follow a step-wise course through successive modes as it did in the FPU case. After some time, the configuration of the system bore little resemblance to the initial state. By that time, modes began to exchange energy more freely and mode transitions at the chain boundary were no longer the dominant influence.

There seemed to be no evidence of the periodic behavior observed by FPU and others. The only significant difference in the energy sharing process between the weak and strong coupling cases was that the rate of mode mixing was greater in the strong coupling than in the weak coupling examples.

We have chosen three figures from the Northcote and Potts paper to exhibit these results. The first, Fig. 16, represents the energy in the first and second modes as a function of collision number. Notice the rapid drop of energy in the lowest mode after the 37th collision, and also notice that most of the energy in the first mode goes directly into modes higher than the second, especially after the 37th collision, without going into the second mode. The constants of the system were chosen to be $N = 15$ particles, $M = 3 \times 10^{-23}$ g., $\gamma = 400$ dynes/cm., $\ell = 4.000 \times 10^{-8}$cm., $a = 3.995 \times 10^{-8}$ cm., and, $d = 3.400 \times 10^{-8}$cm. Figures 17 and 18 compare the manner in which the temperature equilibrates when the chain is initially in the first mode with the energy all initially in the 31st mode. $N = 31$ and Fig. 17 corresponds to $\varepsilon = E/N = 0.7 \times 10^{-14}$ erg (equilibrium temperature $T = 62.0^{\circ}$K), while Fig. 18 corresponds to $\varepsilon = 0.4 \times 10^{-14}$erg and an equilibrium temperature $T = 31.7^{\circ}$K.

Now why does the striking difference exist between the FPU and the Northcote and Potts calculation? Ford's remarks on the importance of resonance effects are irrelevant to the NP calculation because perturbation theory as presented above is not appropriate for the strong hard sphere non-linear model in which the force law cannot be expanded in a power series in the displacements from equilibrium.

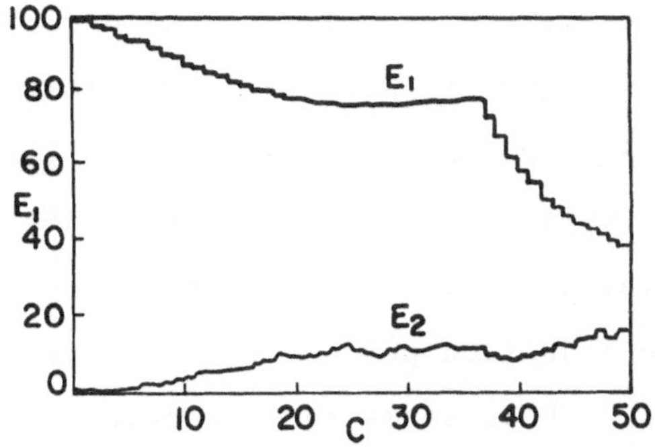

FIGURE 16. The energy in the first two modes after C collisions for N = 15 particles. The mean energy/particle is ε = o.7 x 10^{-14} erg. All energy was initially in the first mode[25].

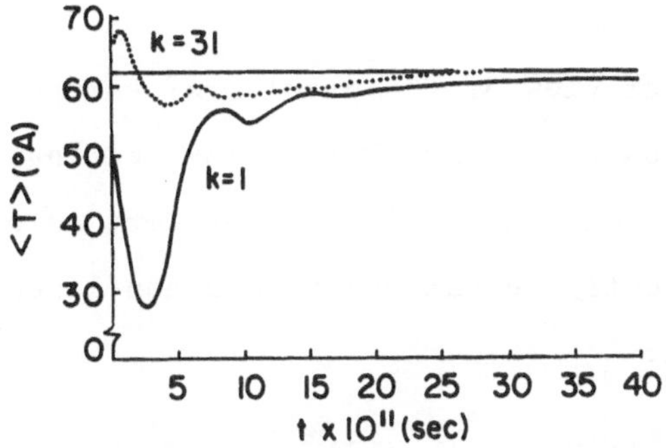

FIGURE 17. The mean temperature computed for the systems N = 31, ε = 0.7 x 10^{-14} erg, and $E_i = \delta_{ik}E$ at t = 0. The predicted thermodynamic temperature is T'' = 62.07°A.

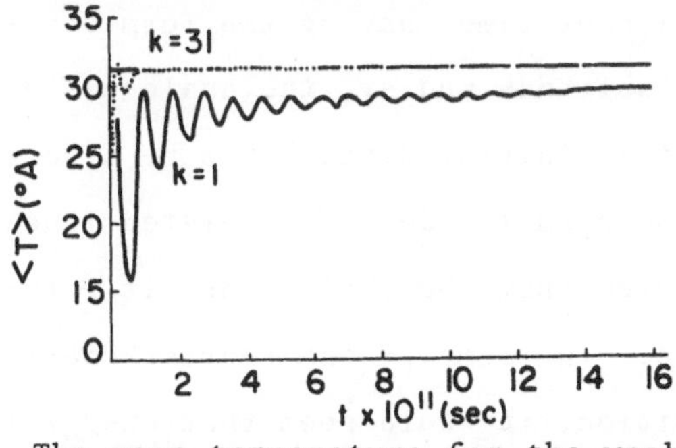

FIGURE 18. The mean temperature for the weakly nonlinear systems N = 31, ε = 0.4 x 10^{-14} erg, $E_i = \delta_{ik}E$ at t = 0. T' = 31.7°A. The distinctly linear behavior of the system for k = 1 is apparent.

The first 2D mode mixing calculations have recently been made by Hirooka and Saito[29] who investigated two dimensional lattices with a quartic anharmonic term in the potential, (i.e., the 2D generalization of [60b]). Computer calculations indicate the existnece of a critical induction period after which the energy shring between modes develops rather rapidly. The induction period increases as the quartic force constant β decreases. There seems to be a critical value of β, say β_0 such that when β exceeds β_0, the system becomes ergodic while, when β is less than β_0, the lattice seems to be almost periodic in the manner exhibited by the FPU 1D calculations.

D. Solitons.

A deeper point of view of the problem of energy transfer between modes was taken by Zabusky and Kruskal[26]. By letting the lattice spacing vanish as was done in Eq. (4b), they were able to convert the FPU difference-differential equations into a non-linear partial differential equation:

$$c^{-2}U_{tt} = U_{xx}(1+\alpha U_x) + (h^2/12)U_{xxxx}+O(h^3) \qquad (5.15)$$

where h is that lattice spacing.

This equation has special solutions which preserve their character as a function of time, having the form $U = f(x-vt)$. These solutions are called solitons and are the analogues of normal modes of linear problems. Let an initial disturbance be a soliton, and let it be decomposed into the normal modes of a related linear problem. When the soliton is reflected from the chain ends, it returns to its initial configuration. In terms of the Fourier coefficients of the normal mode decomposition, it would seem that energy is flowing from one normal mode to another so that when the initial configuration is repeated, the original Fourier components repeat themselves. Zabusky and Kruskal would then explain the FPU results by saying that the initial state of the chain is close to a soliton state which preserves its character for a long time, making it seem that the chain is not ergodic and that energy gets transferred in and out of the lower modes periodically.

We demonstrate here how a soliton solution of (5.15) can be found. We write (5.15) as (with $\gamma^2 = h^2/12$)

$$c^2 U_{tt} = \{U_x + \tfrac{1}{2}\alpha U_x^2 + \gamma^2 U_{xxx}\}_x \tag{5.16}$$

and seek solutions of the form

$$U = f(w) \equiv f(x-vt) \ .$$

Then

$$d/dw\{[1-(v/c)^2]U_w + \tfrac{1}{2}\alpha U_w^2 + \gamma^2 U_{www}\} = 0$$

so that

$$[1-(v/c)^2]U_w + \tfrac{1}{2}\alpha U_w^2 + \gamma^2 U_{www} = C = \text{constant} \ .$$

Now define a new dependent variable W by

$$U = W + \mu w$$

where the constant μ is to be chosen so that

$$\mu^2 + \mu[1-(v/c)^2] - C = 0.$$

Then

$$\{\alpha\mu + [1-(v/c)^2]\}W_w + \tfrac{1}{2}\alpha W_w^2 + \gamma^2 W_{www} = 0.$$

Now let

$$dW/dw = Z,$$

then

$$\gamma^2 Z_{ww} = \lambda^2 Z - \tfrac{1}{2}\alpha Z^2. \tag{5.17}$$

The general solution of this ordinary nonlinear equation is expressible in terms of elliptic integrals. A special solution of some interest is, with A being a constant of integration,

$$Z(w) = (3\lambda^2/\alpha)\operatorname{sech}^2(\lambda w/2\gamma) + A \quad . \tag{5.18}$$

Hence, if B is another constant of integration,

$$U(w) = (3\lambda\gamma/2)\tanh(\lambda w/2\gamma) + (A+\mu)w+B \tag{5.19}$$

which is a soliton solution such that, with reflecting boundary conditions, the characteristics alluded to at the beginning of this section would be observed.

Zabusky's[30] detailed analysis (including numerical work) is based on the connection between (5.15) and the Korteweg-de Fries equation[32]

$$u_t + u \cdot u_x + \delta^2 u_{xxx} = 0 \ . \tag{5.20}$$

This equation was first derived in the 1890's in connection with water waves in narrow channels. Solitons were first observed by canal watchers many years ago. It was noted that the bow wave of a canal boat propagated essentially unchanged for great distances behind the boat. The famous Naval architect of the Great Eastern, Scott-Russell, measured the velocity of propagation of the solitons by riding on horseback behind canal boats[34].

We demonstrate the broad existence of solitons by constructing them for a rather general class of nonlinear wave equations.

The continuum wave equation

$$u_{tt}/c^2 = u_{xx} \tag{5.21}$$

in an unbounded medium has the general solution

$$u = f(x \pm ct) \tag{5.22}$$

so that if the form f(x) of a wave is given at time t = 0, its form is forever the same and its propagation velocity is c.

On the other hand, if we have a nonlinear wave equation such as

$$u_{tt}/c^2 = [F(u)]_{xx} \tag{5.23}$$

then we would expect that mode coupling would develop. One can, however, construct special persistent waves or solitons for any function F(u).

We remember that any function u(w) with $w \equiv x \pm ct$ is a solution of the linear equation (1). Now consider the equation with constants c_1 and c_2:

$$F\{u(w)\} = u(w) + c_1 + c_2 w \quad . \tag{5.24}$$

For any function F(u) one can solve this (perhaps transcendental) equation for u(w). The u(w) may not be a function which interests us, but, nevertheless, one could generally find a solution. If u(w) is such a solution, then (5.24) can be substituted into (5.23) to find (since $\partial^2 w/\partial x^2 = 0$) that $u = u(w)$ is a solution of (5.23).

Let us construct some examples by working backward. Suppose we are interested in a soliton that looks like a shock wave with

$$u(w) = \tfrac{1}{2}(1 + \tanh w) \quad . \tag{5.25a}$$

Then

$$\tanh^{-1}(2u - 1) = w \tag{5.25b}$$

and

$$F(u) = u + c_1 + c_2 \tanh^{-1}(2u-1) \tag{5.25c}$$

so that the non-linear wave equation which has our shock wave type of soliton solution is

$$u_{tt}/c^2 = u_{xx} + c_2\{\tanh^{-1}(2u - 1)\}_{xx} . \tag{5.26}$$

The Gaussian soliton

$$u(w) = \exp(-w^2/a) \tag{5.27a}$$

implies that

$$w = \{a \log(1/u)\}^{1/2} \tag{5.27b}$$

so

$$F(u) = u + c_1 + c_2\{a \log(1/u)\}^{1/2} \tag{5.27c}$$

and

$$u_{tt}/c^2 = u_{xx} + c_2[\{a \log(1/u)\}^{1/2}]_{xx}. \tag{5.27d}$$

Other examples can be constructed.

E. Solitons on Discrete Lattices.

It is not so easy to construct solitons for wave propagation in non-linear discrete lattices. An elegant example, however, was given by Toda[27]. He considered a chain of atoms whose interaction potential was defined by

$$\begin{aligned}\phi(R) &= \text{const.} + a(R-D) + (a/b)e^{-b(R-D)} \\ &= \text{const.} + \tfrac{1}{2}ab(R-D)^2 - (ab^2/6)(R-D)^3 + \ldots\end{aligned} \tag{5.28a}$$

Hence, if ab = const. while $a \to \infty$ and $b \to 0$, then $\phi(R)$ becomes the harmonic potential. However, if $b \to \infty$ for fixed ab,

$$\phi(R) \to \begin{cases} \infty & \text{if } R < D \\ 0 & \text{if } R > D \end{cases} \tag{5.28b}$$

which is the defining characteristic of a hard sphere repulsion.

Now consider a chain of atoms with deviations from equilibrium positions $u_1, u_2, \ldots$. Then the kinetic and potential energies of the chain are

$$T = \Sigma\, p_n^2/2m \qquad \text{with} \qquad p_n = m\dot{u}_n \tag{5.29a}$$

$$\Phi = \Sigma\, \phi(u_n - u_{n-1}) \tag{5.29b}$$

The equations of motion are, in the usual form,

$$\dot{p}_n = m\ddot{u}_n = -\partial\phi(u_n - u_{n-1})/\partial u_n - \partial\phi(u_{n+1} - u_n)/\partial u_n \; . \qquad (5.30a)$$

If we use the interaction law (5.24)

$$m\ddot{u}_n = a\exp\{-b[u_n - u_{n-1} - D]\} - a\exp\{-b[u_{n+1} - u_n - D]\}. \qquad (5.30b)$$

Let us now subtract the relative coordinates

$$r_n = u_n - u_{n-1} \qquad (5.31)$$

into (5.30a). Then

$$m\ddot{u}_n = f(r_n) - f(r_{n+1}) \qquad (5.32a)$$

so that by subtracting the $(n-1)^{st}$ of the equations from the nth, we find

$$-\frac{d}{dt}(\dot{r}_n) = \frac{1}{m}\{f(r_{n+1}) - 2f(r_n) + f(r_{n-1})\} \qquad (5.32b)$$

where now

$$u_1 = r_1;\; u_2 = r_1 + r_2;\; u_3 = r_1 + r_2 + r_3;\; \text{etc.} \qquad (5.33)$$

Now define

$$\dot{s}_n = \frac{1}{m} f(r_n) \qquad (5.34a)$$

and suppose this expression can be inverted so that

$$r_n = \chi(\dot{s}_n) \quad \text{and} \quad \dot{r}_n = \ddot{s}_n \chi'(\dot{s}_n) \; . \qquad (5.34b)$$

Then (5.32) is equivalent to

$$-\ddot{s}_n \chi'(\dot{s}_n) = s_{n+1} - 2s_n + s_{n-1} \qquad (5.35)$$

The specific choice

$$f(r) = -\alpha(1 - e^{-br}) \quad \text{with} \quad \alpha = ae^{bD} \qquad (5.36)$$

yields the connection between (5.32) and (5.35). Then

$$b(s_{n+1} - 2s_n + s_{n-1}) = \ddot{s}_n/(\dot{s}_n + \alpha/m). \tag{5.37}$$

Toda[27] found a special soliton solution of these equations by noticing that this formula resembles one which can be derived from the addition formula for the elliptic function sn u. The following elliptic function identity is well known[28]:

$$\mathrm{sn}^2(u+v) - \mathrm{sn}^2(u-v) = 2\frac{d}{dv}\left\{\frac{\mathrm{sn}\,u\ \mathrm{cn}\,u\ \mathrm{dn}\,u\ \mathrm{sn}^2 v}{1 - k^2\mathrm{sn}^2 u\ \mathrm{sn}^2 v}\right\} \tag{5.38}$$

where k is the molulus of the Jacobi elliptic functions and

$$\mathrm{dn}^2 u = 1 - k^2\mathrm{sn}^2 u. \tag{5.39a}$$

Then if one defines

$$\varepsilon(n) = \int_0^u \mathrm{dn}^2 u\ dn, \tag{5.39b}$$

$$\varepsilon'(u) = \mathrm{dn}^2 u \text{ and } \varepsilon''(u) = -2k^2\mathrm{sn}\,u\ \mathrm{cn}\,u\ \mathrm{dn}\,u \tag{5.39c}$$

so that

$$k^2\mathrm{dn}^2(u+v) - k^2\mathrm{dn}^2(u-v) = -\frac{d}{dv}\left\{\frac{\varepsilon''(u)}{\mathrm{sn}^{-2}v - 1 + \varepsilon'(u)}\right\}, \tag{5.40a}$$

or

$$\int_0^v \mathrm{dn}^2(u-v)dv - \int_0^v \mathrm{dn}^2(n+v)dv = -\frac{\varepsilon''(u)}{\mathrm{sn}^{-2}v - 1 + \varepsilon'(u)}. \tag{5.40b}$$

Then

$$\varepsilon(u+v) - 2\varepsilon(u) + \varepsilon(u-v) = \frac{\varepsilon''(u)}{-1 + \mathrm{sn}^{-2}v + \varepsilon'(u)}. \tag{5.41}$$

The structure of this formula closely resembles that of (5.37). It is known that[28]

$$Z(u) \equiv \varepsilon(u) - uE/K = \frac{\partial}{\partial u}\log\Theta_4(u/2K) \tag{5.42}$$

where E and K are, respectively, complete elliptic integrals of the first and second kind of the variable k, and Θ_4 is the fourth theta function. Then

$$Z(u+v) - 2Z(u) + Z(u-v) = \frac{Z''(u)}{-1 + sn^{-2}v + (E/K + Z'(u))} . \qquad (5.43)$$

The reader, by direct differention, can verify that not only do (5.41) and (5.37) resemble each other, but that a special solution of (5.37) is

$$s_n(t) = (2K\nu/b)\, Z\{2K(\nu t \pm n\lambda)\} \qquad (5.44)$$

as seen by setting

$$v = 2K/\lambda, \qquad u = 2K(\nu t \pm n/\lambda) \qquad (5.45a)$$

and observing the dispersion relation between λ and ν

$$-1 + sn^{-2}(2K/\lambda) + (E/K) = \frac{\alpha b}{m(2K\nu)^2}$$

or

$$2K\nu = \{(\alpha b/m)/[-1 + sn^{-2}(2K/\lambda) + E/K]\}^{1/2} . \qquad (5.45b)$$

The function $Z(u)$ is periodic with $(Z(u+2K) = Z(u)$ and ν and λ are, respectively, the frequency and wavelength of the soliton. Finally, by combining (5.34b) and (5.44)

$$r_n = -\frac{1}{b} \log\{1+\frac{4(K\nu)^2 m}{\alpha b} [dn^2\{2K(\nu t \pm n/\lambda)\}-\frac{E}{K}]\} \qquad (5.46)$$

when the modulus k is very small

$$sn\ u \simeq \sin u;\ E/K \simeq 1 - \tfrac{1}{2}k^2,\ Z(u) \simeq (K^2/4)\sin 2u,\ K \simeq \pi/2,$$

and if $\gamma = \alpha b$

$$s_n \simeq \frac{\omega k^2}{8b} \sin(\omega t \pm 2\pi n/\lambda) \text{ with } \omega = 2(\gamma/m)^{1/2}\sin\pi/\lambda .$$

Furthermore

$$r_n \simeq -(\omega^2 k^2/8\alpha b^2)\cos(\omega t \pm 2\pi n/\lambda) \qquad (5.47)$$

which corresponds to a typical wave which propagates in the harmonic lattice.

The function which appears in the soliton formulae,

$$dn^2(2Kx) - (E/K)$$

is plotted in Figure 19. As $k \to 1$ with $u - 2K/\lambda$ fixed, the various elliptic functions reduce to hyperbolic ones.

While all continuum wave equations of the type (5.23) have soliton solutions, it seems that more conditions must be satisfied for discrete wave equations. I suspect that the model used by Northcote and Potts does not have one.

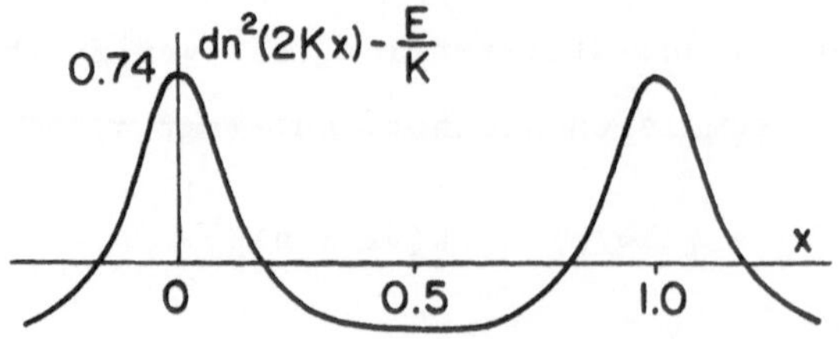

FIGURE 19. $f(x) = dn^2(2kx) - E/K$ as a function of x, for $k^2 = 0.992$ where k is the modulus[27].

F. Exactly Solvable Wave Equations.

While the Toda model is very ingenious and leads to soliton solutions, it has not been possible to find a general solution for an arbitrary set of initial conditions. We now introduce two classes of exactly solvable nonlinear models, one a nonlinear wave equation and the other, while it looks like a rate process, is really a wave propagation process.

First we shall consider a faltung type integro-differential equations

$$2\frac{d^2U(r,t)}{dt^2} = 12\int_{-\infty}^{\infty} S_2(r-r')U(r',t)dr' + 12\iint_{-\infty}^{\infty} S_3(r-r'-r'')U(r',t)U(r'',t)dr'dr''$$
$$+4\iiint_{-\infty}^{\infty} S_4(r-r'-r''-r''')U(r',t)U(r'',t)U(r''',t)dr'dr''dr''' + 4S_1(r) \quad (5.48a)$$

There is a corresponding lattice model with N lattice points and periodic boundary conditions such that $n = 0,1,2,\ldots,N-1$. Then

$$2\frac{d^2U(r,t)}{dt^2} = 12\sum_{n'=0}^{N-1} S_2(n-n')U(n',t) + \ldots + 4S_1(r). \quad (5.48b)$$

There is no difficulty in constructing corresponding 2 and 3 dimensional models.

An analysis of the continuous equation employs the Fourier transforms

$$\begin{Bmatrix} u(k,t) \\ s_j(k) \end{Bmatrix} = \int_{-\infty}^{\infty} \begin{Bmatrix} U(r,t) \\ S_j(r) \end{Bmatrix} \exp(ikr)dr \quad (5.49a)$$

while that for the discrete case employs finite Fourier expansions

$$\begin{Bmatrix} u(2\pi\ell/N) \\ s_j(2\pi\ell/N) \end{Bmatrix} = \sum_{n=0}^{N-1} \begin{Bmatrix} U(n,t) \\ S_j(n,t) \end{Bmatrix} \exp(2\pi i\ell n/N) . \tag{5.49b}$$

An application of the Faltung theorem in either case yields

$$2\,\frac{d^2u}{dt^2} = 4s_1 + 12s_2u + 12s_3u^2 + 4s_4u^2 . \tag{5.50}$$

Now multiply by ½ (du/dt) and integrate. Then we have as the first integral (where s_0 is the constant of integration)

$$\begin{aligned} &(du/dt)^2 = s_0 + 4s_1u + 6s_2u^2 + 4s_3u^3 + s_4u^4 \equiv f(u) \\ &\int_{u(t_0)}^{u(t)} du[f(u)]^{-1/2} = t-t_0 . \end{aligned} \tag{5.51}$$

This equation can be solved for u(t) as was first done by Weierstrass (cf. Chapter XX in Whitaker and Watson [28]). Let

$$p(z) \equiv p(z,g_2,g_3) \tag{5.52}$$

be the elliptic function characterized by the invariants of f(u),

$$g_2 = s_0s_4 - 4s_3s_1 + 3s_2^2 \tag{5.53}$$

$$g_3 = s_0s_2s_4 + 3s_1s_2s_3 - s_2^3 - s_4s_1^3 - s_0s_3^3 . \tag{5.54}$$

One finds that [28] if $u(t_0) \equiv u_0$

$$u(t) - u(t_0) \equiv P(t-t_0,k) \equiv$$

$$\frac{[f(u_0)]^{1/2}p'(t-t_0') + \frac{1}{2}f'(u_0)\{p(t-t_0) - \frac{1}{24}f''(u_0)\} + \frac{1}{24}f(u_0)f'''(u_0)}{2\{p(t-t_0) - \frac{1}{24}f''(u_0)\}^2 - \frac{1}{48}f(u_0)f^{iv}(u_0)} . \tag{5.55}$$

Fourier inversion yields

$$U(r,t) = U(r,t_0) + (1/2\pi) \int_{-\infty}^{\infty} P(t-t_0,k)e^{-ikr}\,dk \qquad (5.56)$$

Corresponding expressions exist for discrete and 3D cases.

If the linear approximation to the right hand side of (5.48a) is to have the same form as the standard wave equation, one chooses

$$s_2(r) \equiv (1/6)C^2\delta''(r) \; . \qquad (5.57)$$

The integration by parts

$$\int s_2(r-r')U(r')\,dr' = (1/6)C^2 \int \delta''(r-r')U(r')\,dr'$$

$$= (1/6)C^2\partial^2U/\partial r^2 \; .$$

In the lattice case, an interesting choice of $s_j(n)$ is

$$s_2(n) = 1/6\gamma\{\delta(n-1) - 2\delta(n) + \delta(n+1)\}, \qquad (5.58a)$$

$$s_1 = 0 \; , \qquad s_4 = 0 \; , \qquad (5.58b)$$

$$s_3(n) = 1/6\alpha\{\delta(n-1) - 2\delta(n) + \delta(n+1)\}. \qquad (5.58c)$$

Then

$$\frac{d^2U(n)}{dt^2} = \gamma\{U(n-1) -2U(n) + U(n+1)\}$$
$$+ \alpha \sum_{k'} U(n')\{U(n-n'+1) -2U(n-n') + U(n-n'-1)\}. \qquad (5.59)$$

Since this equation can be solved exactly, it can be used to investigate mode coupling and can be extended to the two and three dimensional analogues. An interesting feature of this model is that if one wishes to follow the distribution of energy in normal modes of the unperturbed problem, the inversion (5.56) does not have to be carried out. A detailed discussion of this model will be given elsewhere.

The second model which we consider is one suggested by the Volterra prey-predator equations. They are of the form

$$dN_i/dt = k_i' N_i + \sum_{j=1}^{n} a_{ij} N_i N_j \quad . \tag{5.60a}$$

The first term describes the behavior of the ith species in the absence of others; when $k_i' > 0$, the ith species is postulated to grow in an exponential Malthusian manner. When $k_i' < 0$, the ith species dies out exponentially in the absence of prey. Here one sets

$$a_{ij} = -a_{ji} \tag{5.60b}$$

since, if i is eaten by j, the binary interaction term is negative for the i species and positive for the jth. Equation (5.60) might alternatively be written as

$$dN_i/dt = k_i N_i + \sum_j a_{ij} N_i (N_j - 1) \tag{5.60c}$$

where

$$k_i' = k_i - \sum_j a_{ij} N_i \quad . \tag{5.60d}$$

If (5.60c) is taken to be basic, it is the member of a general class of equations

$$dN_i/dt = k_i N_i^{\gamma_i} + \sum_{j=1}^{n} a_{ij} N_i^{\gamma_i} (N_j^{\alpha_j} - 1)/\alpha_j \tag{5.61}$$

with $\gamma_i = \alpha_i = 1$ for all i. When $\gamma_i = 1 - \alpha_i$, the general set can be written

$$\alpha_i^{-1} d(N_i^{\alpha_i} - q_i^{\alpha_i})/\alpha_i = \sum a_{ij} [N_j^{\alpha_j} - q_j^{\alpha_j}]/\alpha_j \tag{5.62}$$

where $(q_j^{\alpha_j} - 1)/\alpha_j$ is the solution of the linear equilibrium equation characterized by $dN_i/dt = 0$ for all i)

$$\sum_j a_{ij} (q_j^{\alpha_j} - 1)/\alpha_j = -k_i \quad . \qquad (5.63$$

The attractive feature of the nonlinear set (5.62) is that, if we let

$$u_i = (N_i^{\alpha_i} - q_i^{\alpha_i})/\alpha_i \tag{5.64}$$

then the set (5.62) is linear in the new variables

$$du_i/dt = \sum_j a_{ij} u_j \tag{5.65}$$

or in matrix form, with A being anti-symmetrical),

$$du/dt = Au \tag{5.66a}$$

so that

$$d^2u/dt^2 = A^2 u \ . \tag{5.66b}$$

Since the square of an anti-symmetric matrix is symmetrical, we expect oscillatory solutions of the type one finds for wave equations, rather than the damped solutions of the type observed for rate equations. It is the anti-symmetry of A which is responsible for this behavior.

In the special case $\alpha_i = 0$ for all i,

$$u_i = \log(N_i/q_i) \ . \tag{5.67}$$

If we consider a one-dimensional ring of 2N species with

$$a_{ij} = a(\delta_{j,i+1} - \delta_{j,i-1}) \text{ and } u_j = u_{j+2N} \tag{5.68}$$

then

$$du_j/dt = a(u_{j+1} - u_{j-1}) \tag{5.69}$$

whose normal mode solutions are

$$u_j = u \sin(j\theta_k \pm \omega_k t) \tag{5.70a}$$

with

$$\theta_k = 2\pi k/2N \qquad \text{and} \qquad \omega_k = 2a \sin\theta_k . \tag{5.70b}$$

The + sign corresponds to a wave which propagates in the negative direction while the - sign refers to one which propagates in the positive direction.

The sum of the functions which represent waves going in opposite directions is also a solution of the linear equations. Such a solution is

$$v_j = k[\sin(j\theta_k + \omega_k\tau) + \sin(j\theta_k - \omega_k\tau)] \qquad (5.71)$$

where k represents the amplitude of the waves. When this is transformed to the population variables, one finds

$$N_j(\tau) = q_j \exp k\{\sin(j\theta_k + \omega_k\tau)\} \exp k\{\sin(j\theta_k - \omega_k\tau)\} . \qquad (5.72)$$

In the regime of large k, the sine wave solution of the linear equation is sharpened for those values of $(j\tau_k \pm \omega_k\tau) > 0$ and flattened when $(j\theta \pm \omega_k\tau) < 0$. The exponentials are plotted in Fig. 20 with $k = 10$.

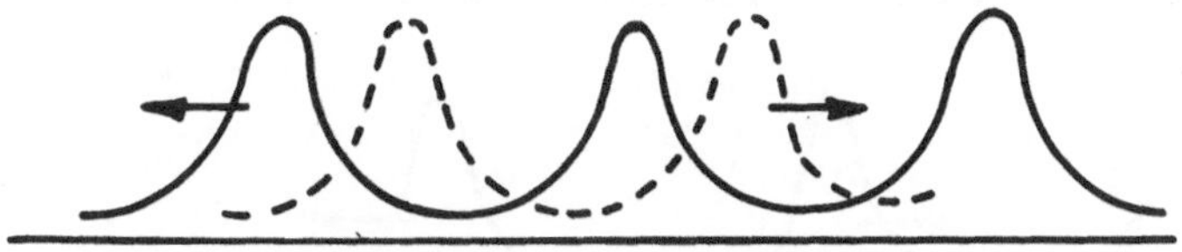

FIGURE 20. Form of soliton wave solution of our nonlinear rate equations.

When the two exponentials are out of phase by 180°, the product of the peak of one and the trough of the other is 1, while in a regime in which two troughs coincide, the product is e^{-20}; it has the value e^{20} when both factors are in phase. We consider long wave length solitons in Fig. 21 and note the behavior of two peaks "a" and "b" which are moving toward each other. When they collide, the variable v_j goes wild, changing its value by a factor of e^{20}. After the collision both solitons continue on their way as though the collision never occurred. The solitons preserve their character in a manner analogous to that of stable particles in nature.

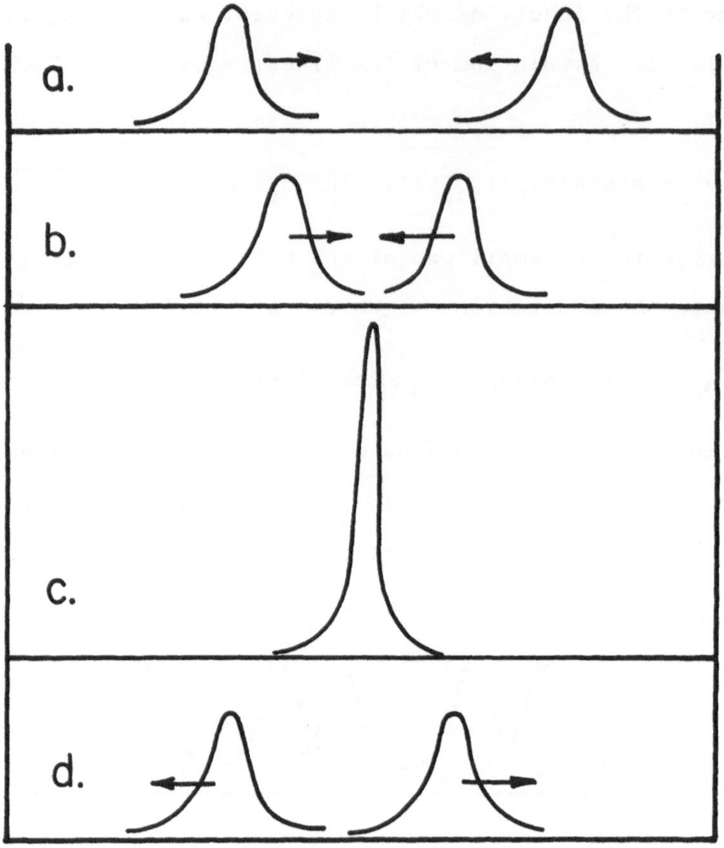

FIGURE 21. When our two solitons approach each other as in (a), they eventually overlap as in (c) with tremendous distortion. Then they emerge from the entanglement in their original form as in (d).

ACKNOWLEDGEMENTS

Part of this lecture was published in the "Proceedings of the Explosive Chemical Reactions Seminar" of 21-23 October, 1968, which was sponsored by the U. S. Army Research Office and U. S. Army Ballistic Research Laboratories. Part of the research has been sponsored by ARPA and monitored by ONR under Contract No. N00014-67-0398-0005.

References:

[1] cf. L. BRILLOUIN: Wave Propagation (McGraw-Hill, 1946; or Dover Reprint, 1953).

[2] E. SCHRODINGER: Ann. Phys. 44, 191 (1914).

[3] P. M. MORSE AND K. V. INGURD: Theoretical Acoustics (see, explicitly, p. 89-91)(McGraw-Hill, 1968).

[4] J. J. GILMAN AND G. H. VINEYARD: Preprint.

[5] cf. A. MARADUDIN, E. MONTROLL AND G. WEISS: Theory of Lattice Vibrations in the Harmonic Approximation (Academic Press, 1963).

[6] W. A. BOWERS AND H. B. ROSENSTOCK: J. Chem. Phys. 18, 1056 (1950).

[7] H. B. ROSENSTOCK AND G. F. NEWELL: J. Chem. Phys. 21, 1607 (1953).

[8] E. W. MONTROLL: Proc. of 3rd Berkeley Symposium on Mathematical Statistics and Probability, Vol. III, p. 209, 1956.

[9] E. MONTROLL: J. Chem. Phys. 15, 575 (1947); Am. Math. Monthly LXI, 46 (1954).

[10] L. VAN HOVE: Phys. Rev. 89, 1189 (1953).

[11] H. B. ROSENSTOCK: Phys. Rev. 97, 290 (1955).

[12] J. C. PHILLIPS: Phys. Rev. 104, 1263 (1956).

[13] cf. W. COCHRAN: Reports on Progress in Physics 26, 1 (1963), and B. N. BROCKHOUSE AND A. T. STEWART: Rev. Mod. Phys. 30, 236 (1958).

[14] A. MARADUDIN, PAUL MAZUR, E. MONTROLL AND G. WEISS: Rev. Mod. Phys. 30, 175 (1958).

[15] D. N. PAYTON III, M. RICH, AND W. M. VISSCHER: Localized Excitations in Solids, p. 657 (Plenum, 1968).

[16] E. W. MONTROLL AND R. B. POTTS: Phys. Rev. 100, 525 (1955).

[17] I. M. LIFSHITZ: Nuovo Cimento Suppl. A1, 3, 591 (1956).

[18] A. MARADUDIN: Reports on Progress in Physics 28, 331 (1965).

[19] I. M. LIFSHITZ AND A. M. KOSEVICH: Reports on Progress in Physics 29, 216 (1966).

[20] R. WALLIS (editor): Localized Excitations in Solids (Plenum, 1968).

[21] E. FERMI, J. PASTA AND S. ULAM: Collected Works of Enrico Fermi, Vol. II, p. 978 (Chicago, 1965).

[22] J. FORD: J. Math. Phys. 2, 387 (1961); J. FORD AND J. WATERS: J. Math. Phys. 4, 1293 (1963).

[23] E. A. JACKSON: J. Math. Phys. 4, 551 (1963).

[24] E. W. BROWN: Bull. Am. Math. Soc. 34, 265 (1928).

[25] R. S. NORTHCOTE AND R. B. POTTS: J. Math. Phys. 5, 383 (1964).

[26] N. J. ZABUSKY AND M. D. KRUSKAL: Phys. Rev. Letters 15 241 (1965).

[27] M. TODA: J. Phys. Soc. Japan 22, 431 (1967); 23, 501 (1968).

[28] E. T. WHITTAKER AND G. N. WATSON: Modern Analysis, Chapter 22 (Cambridge, 1927).

[29] H. HIROOKA AND N. SAITO: J. of Phys. Soc. of Japan 26 624 (1969).

[30] N. J. ZABUSKY: Proc. of Conference on Math. Models in Phys. Sci., p. 99 (S. DROBOT, editor)(Prentice Hall, 1963).

[31] N. J. ZABUSKY: Proc. Symp. on Nonlinear Partial Differential Equations (Editor, W. AMES)(Academic Press, 1967).

[32] D. J. KORTEWEG AND G. DE VRIES: Phil. Mag. 39, 442 (1895).

[33] T. VON KARMAN: Bull. Am. Math. Soc. 46, 615 (1940).

STOCHASTIC BEHAVIOR IN NONLINEAR OSCILLATOR SYSTEMS*

Joseph Ford
Georgia Institute of Technology
Atlanta, Georgia

CHAPTER I. INTRODUCTION

Despite the fact that Newton's equations have been known for approximately three hundred years, very little information concerning the generic character of their solutions exists. There is, for example, a long standing, if minor, scandal in astronomy concerning the lack of a proof for the stability of the solar system. Closer to home for physicists, the two-body problem lies beyond the reach of present day analysis, except in the special case (of considerable physical interest) for which the center of mass motion separates from the relative particle motion. Even harmonic oscillator systems, which physicists usually assume to be completely solvable systems, can be solved in complete detail only under very severe restrictions. There is, for example, no known general analytic expression for the density of vibrational frequencies of even a one-dimensional, isotopically disordered, harmonic crystal.

Equilibrium statistical mechanics, which chooses to discuss the many-body problem despite our monumental ignorance, sidesteps this issue by using an assumption. It assumes that the system phase space trajectory for an isolated system wanders freely over the energy surface. It then further assumes that the vast majority of micro-states (Q_i, P_i) on the energy surface correspond to macroscopic, thermodynamic equilibrium states (P, V, T, etc.). Thus an isolated system started in some disequilibrium state, wandering freely on the energy surface, will surely tend to equilibrium. The approach to equilibrium might not be monotonic and fluctuations might occur, but the generic pattern is clear. Moreover, in this picture, one clearly obtains equilibrium quantities merely by calculating phase space

*This work supported in part by the National Science Foundation.

averages over the energy surface.

In irreversible statistical mechanics, however, life is much more complicated. While the above generic pattern leading to irreversibility is clear, in order to derive a calculative scheme governing the approach to equilibrium, one must establish an understanding of the basic mechanism causing the wandering phase space trajectories. Now for dilute gas systems, simple two-body collisions yield the wandering trajectories, and the Boltzmann Equation incorporates these two-body collisions into a calculative scheme. For dense systems, it is not yet clear that we fully understand the basic mechanism of wandering trajectories nor that we have a completely adequate calculative scheme. Thus in this paper, using nonlinear oscillator systems as a model, we shall illustrate the source of wandering trajectories for a dense system. The development of the calculative scheme lies in the future.

CHAPTER II. THE MODEL

The Hamiltonian describing oscillatory motion about an equilibrium point may be written

$$H = \sum_{k=1}^{N} (\omega_k/2)(P_k^2 + Q_k^2) + V_3 + V_4 + \ldots, \tag{2.1}$$

where N is the number of oscillators, ω_k are the positive frequencies of the harmonic approximation, V_3, V_4, etc. are cubic, quartic, etc. polynomials in the Q_k, P_k variables. For purposes of this discussion it is convenient to introduce "polar" coordinates via

$$Q_k = (2J_k)^{1/2} \cos\phi_k, \tag{2.2}$$
$$P_k = -(2J_k)^{1/2} \sin\phi_k,$$

where the square root appears in order that the transformation be canonical. Hamiltonian (2.1) may then be written

$$H = H_0(J_1,\ldots,J_N) + V(J_1,\ldots J_N, \phi_1,\ldots,\phi_N), \tag{2.3}$$

where all the pure J-terms have been included in H_0. It perhaps s. ould be emphasized that H_0 in eq. (2.3) is not, in general, merely a linear function of the J's; thus H_0 describes a set of strictly nonlinear oscillators. If the potential V in eq. (2.3) were identically zero, then we would have

$$\dot{J}_k = 0 \text{ and } \dot{\phi}_k = \partial H_0/\partial J_k \equiv \Omega_k(J_1,\ldots,J_N). \tag{2.4}$$

Equation (2.4) yields the solutions

$$J_k = \text{Constants}, \quad \phi_k = \Omega_k t + \phi_{ko}. \tag{2.5}$$

For a two-oscillator system, it is convenient to visualize the system trajectory given by eq. (2.5) as lying on the surface of a torus as shown in Fig. 1.

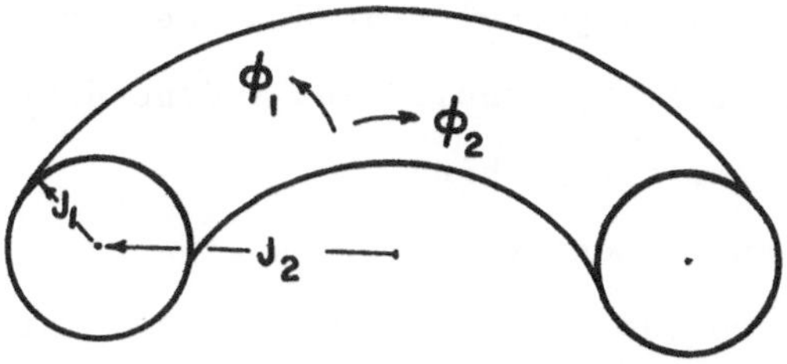

FIGURE 1. A cross-sectional view of a two-dimensional torus showing the coordinates J_1, J_2, ϕ_1, ϕ_2 on the torus.

Here J_1 and J_2 are the "radii" of the torus and ϕ_1 and ϕ_2 are the angle variables on the torus. Ω_1 and Ω_2 give the frequencies of the (generally) quasi-periodic motion on the torus. When N, the number of oscillators, is greater than two, it is difficult to draw the tori which bear the (generally) quasi-periodic motion described by eq. (2.5), but the generalization is nonetheless straightforward. Regardless, when V is identically equal to zero in eq. (2.3), all system trajectories are seen to lie on smooth, N-dimensional integral surfaces (called tori) embedded in the 2N-dimensional phase space. Thus the case $V \equiv 0$ is quite similar to the case of harmonic systems with the exception that here the frequencies Ω_1 and Ω_2 depend on the constant J's. But most important when $V \equiv 0$, the system trajectory does not wander freely over the energy surface. Nonetheless statistical mechanics assumes that when V is not identically zero, even though very small, then the system trajectory does wander freely over the (2N-1)-dimensional energy surface. In short, statistical mechanics assumes that a little nonlinearity goes a long way!

However in 1954 Kolmogorov[1] stated a theorem, whose proof was published a decade later by Arnold[2] and independently by Moser[3], which asserts that if, among other things,

i) V is sufficiently small, and

ii) the Jacobian of the frequencies Ω_k with respect to the J_k is not identically zero,

then most system trajectories continue to lie on smooth N-dimensional integral surfaces (tori) embedded in the 2N-dimensional phase space. Indeed these perturbed integral surfaces are only slightly distorted versions of the $V \equiv 0$ tori. Thus the Kolmogorov-Arnold-Moser Theorem (hereafter referred to as KAM) leaves statistical mechanics with something of a problem.

The key to the resolution of the difficulty is that "most trajectories" loophole. Why only "most trajectories"? KAM prove their theorem by changing coordinates from the (J_k, ϕ_k) set to a $(\mathcal{J}_k, \theta_k)$ set such that Hamiltonian (2.3), written as

$$H = H_o(J_1,\ldots,J_N) + \ldots + f_{\{n_k\}}\cos(\Sigma n_k\phi_k) + \ldots, \tag{2.6}$$

where the V of eq. (2.3) has been expanded in a Fourier series and only one typical term has been explicitly written out in eq. (2.6), becomes

$$\mathcal{H} = \mathcal{H}(\mathcal{J}_1,\ldots,\mathcal{J}_N), \tag{2.7}$$

where $\mathcal{H}$ is a function of the new momentum coordinates $\mathcal{J}_k$ alone. Clearly if a well-behaved (analytic) canonical transformation could be found which carries eq. (2.6) into eq. (2.7), then all the system trajectories would lie on tori; however, KAM assert only that this transformation exists for most initial conditions. In order to clarify the matter, let us write out the first two terms in one of the transformation equations which would be used to eliminate the specific angle dependent term appearing in eq. (2.6). Though it may not be obvious, the desired equation reads

$$J_1 = J_1 - \frac{f_{\{n_k\}}\cos(\Sigma n_k\phi_k)}{\Sigma n_k\Omega_k} + \ldots\,. \tag{2.8}$$

Inspection of eq. (2.8) reveals that KAM condition (i) is needed to insure that all $f_{\{n_k\}}$ be small, thereby aiding convergence and insuring that the perturbed tori lie close to the unperturbed. KAM condition (ii) is needed to insure that none of the frequency denominators appearing in eq. (2.8) are identically zero. However, it must be noted that even with condition (ii), as the J_k range over their allowed values, the frequency denominators will still be zero for a countably dense set of (Ω_k) values. Moreover each member (Ω_k) of this dense set will have a finite resonance-width since the Ω_k are functions of the J_k and these latter vary. Thus one is led to wonder how the KAM theorem can be true for any initial conditions, much less for the majority.

A very intuitive argument which clarifies the matter involves showing that countably dense sets have measure zero. Consider the unit interval (0,1) on the real axis, and let us delete not only the rationals (h/k), where k > h, but also an interval

$$\frac{h}{k} - \frac{\varepsilon}{k^3} \le \frac{h}{k} \le \frac{h}{k} + \frac{\varepsilon}{k^3} \tag{2.9}$$

about each rational. The total length of the deleted interval is

$$\sum_{k=1}^{\infty} \sum_{h=1}^{k} \left(\frac{2\varepsilon}{k^3}\right) = 2\varepsilon \sum_{k=1}^{\infty} \frac{1}{k^2}\ , \tag{2.10}$$

which may be made as small as we please since the series converges and we may choose ε as small as we please. In analogy, KAM are forced to exclude from their proof not only those initial conditions lying on the tori bearing commensurate frequencies, since such frequencies lead to zero denominators, but also to exclude initial conditions lying within a resonance zone around each of these tori. Nonetheless in analogy with the above argument, the measure of the excluded initial conditions is small provided that V is small. The remaining, non-dense set of tori which are preserved even in the presence of a non-zero V are those tori that bear incommensurate frequencies which are very poorly approximated by the rationals.

From the view point of statistical mechanics, however, we must focus our attention on that small set of trajectories which do not lie on preserved tori and inquire about the character of such trajectories. Returning to eq. (2.8), let us note that if an angle dependent term $\cos(\Sigma n_k \phi_k)$ yields a zero or near zero denominator $\Sigma n_k \Omega_k$ for some band of J_k-values, we must anticipate, in general, that there will be a host of other angle dependent terms $\cos(\Sigma n_k' \phi_k)$ such that $\Sigma n_k' \Omega_k \simeq 0$ for some J_k-values lying in the same band. As a consequence, trajectories in such a region are simultaneously affected by many overlapping resonant interactions. Thus one has here the situation envisioned in the quantum mechanical Golden Rule[4] in which an initial state $(J_1, \ldots, J_N)$ is resonantly coupled to a density of final states $(J_1', \ldots, J_N')$, leading to irreversible behavior. Thus in regions of destroyed tori, one anticipates that the trajectories are very erratic indeed, perhaps ergodicly filling the destroyed regions[5]. Thus if we can only increase the size of this relatively small destroyed region, statistical mechanics can proceed as the founding fathers suggested. In the next two chapters, we demonstrate by example that violation of either of the KAM assumptions is sufficient for widespread breakdown of preserved tori and for erratic or stochastic[6] behavior of most trajectories.

CHAPTER III. STOCHASTICITY FOR LARGE NONLINEARITY

In order to illustrate the freely wandering trajectories which arise in the presence of large nonlinearities, we need only consider a two-oscillator system. In a paper on statistical mechanics it may appear strange indeed to see a detailed discussion of various two-oscillator systems; but we are interested here only in illustrating the source of irreversibility, and, as we shall see, a two-oscillator system exhibits much of the complexity of the many-body problem. Since in Chapter II we argued that resonance plays a central role in the development of stochastic trajectories, let us begin by considering the effect of an isolated, single resonant interaction. In particular let us consider

$$H = J_1 + J_2 - J_1^2 - 3J_1J_2 + J_2^2 + \alpha J_1J_2\cos(2\phi_1 - 2\phi_2), \tag{3.1}$$

where $H_o = J_1 + J_2 - J_1^2 - 3J_1J_2 + J_2^2$. It is clear from eq. (2.8) that the cos-term represents a resonant interaction in that region of phase space for which $2\Omega_1 \cong 2\Omega_2$. In the language of solid state physics, the cos-term represents the resonant, four-phonon interaction $2\Omega_1 \rightleftarrows 2\Omega_2$. Indeed since

$$\Omega_1 \equiv \partial H_o/\partial J_1 = 1 - 2J_1 - 3J_2, \tag{3.2a}$$

and

$$\Omega_2 \equiv \partial H_o/\partial J_2 = 1 - 3J_1, + 2J_2, \tag{3.2b}$$

the cos-term would be expected to strongly distort the unperturbed tori lying in the region for which $J_1 \cong 5J_2$. These assertions could be rigorously proved for Hamiltonian (3.1) since (J_1+J_2) is a constant of the motion in addition to the Hamiltonian which means that the equations of motion for this system can be solved analytically; however, we choose to proceed along other lines.

We would like to determine the generic character of trajectories generated by a specified two-oscillator Hamiltonian. Are they generally stochastic or do most lie on tori? To this end, let us suppose that the Hamiltonian, which is a constant of the motion, has been written $H = H(Q_1, P_1, Q_2, P_2)$ using eq. (2.2) if necessary. Now we may plot the system trajectory in only three dimensions, using (Q_1, P_1, Q_2) say, since the fourth coordinate, here P_2, can be determined from the constant Hamiltonian H. In particular, if the system trajectory, plotted in the (Q_1, P_1, Q_2) space, lies on a two-dimensional surface (torus), then the intersection points of this trajectory, with say the (Q_1, P_1) plane, will lie on a curve. On the other hand if the trajectory is stochastic, then its intersection points with the (Q_1, P_1) plane will form a collection of randomly scattered points. Moreover, since we shall only consider the bounded motion of these Hamiltonian systems, all intersection points for fixed H must lie on or within some bounding curve. Thus we can present a complete graphical picture of all allowed trajectories for a two-oscillator Hamiltonian by plotting trajectory intersections with the (Q_1, P_1) plane for a representative sampling of trajectories. In particular, we may numerically integrate the equations of motion on a computer and determine the (Q_1, P_1) points on each trajectory for which $Q_2 = 0$; a plot of these points in the (Q_1, P_1) plane will then reveal the character of the trajectory.

For Hamiltonian (3.1), a typical plot of trajectory intersections with the (Q_1, P_1) plane is presented in Fig. 2. Were $\alpha = 0$ in Hamiltonian (3.1), each trajectory would yield a circle (centered on the origin) of intersection points in the (Q_1, P_1) plane, since $J_1 = (\frac{1}{2})(Q_1^2 + P_1^2)$ would then be a constant of the motion.

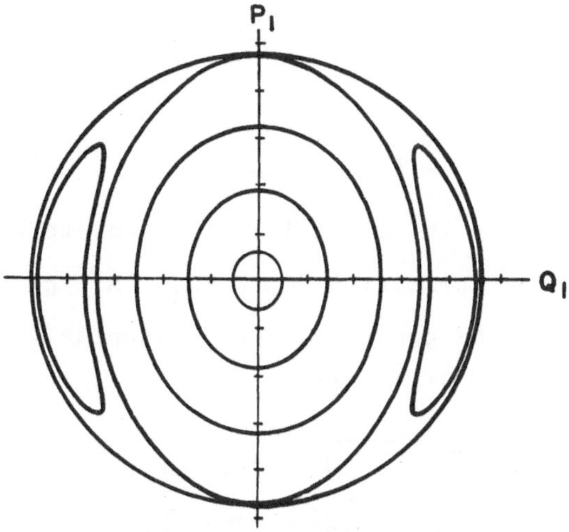

FIGURE 2. Curves of trajectory intersections with the (Q_1,P_1) plane for Hamiltonian (3.1). Whenever the trajectory intersection points appear to lie on a smooth curve, the curve is drawn in free hand.

In a sense, these $\alpha = 0$ circles represent cross-sections of the unperturbed tori upon which each trajectory must lie. When $\alpha \neq 0$, most of the $\alpha = 0$ tori are only slightly deformed, their cross-sections changing to ovals. However those tori lying in the region $J_1 = 5J_2$ where $2\Omega_1 = 2\Omega_2$ are grossly distorted, their cross-sections changing from circles centered on the origin to crescents centered about either of two stable periodic orbits. The isolated points representing these periodic orbits lie on the Q_1-axis. The dividing curve between the crescent and oval regions is called a separatrix, and the self-intersection points of this curve represent two unstable periodic orbits. Trajectories on the torus yielding this separatrix curve are asymptotic to one or another of the unstable periodic orbits. Finally, since $(J_1 + J_2)$ is a second constant of the motion for Hamiltonian (3.1) one may rigorously show[7,8] that all trajectories lie on tori and yield smooth curves in the (Q_1,P_1) plane.

Next let us consider the near-by, isolated, resonant interaction given by

$$H = J_1 + J_2 - J_1^2 - 3J_1J_2 + J_2^2 + \lambda J_1 J_2^{3/2}\cos(2\phi_1 - 3\phi_2), \qquad (3.3)$$

where H_0 is the same as in eq. (3.1). This cos-term represents the resonant five-phonon interaction $2\Omega_1 \rightleftarrows 3\Omega_2$. A typical plot of trajectory intersections generated by Hamiltonian (3.3) appears in Fig. 3.

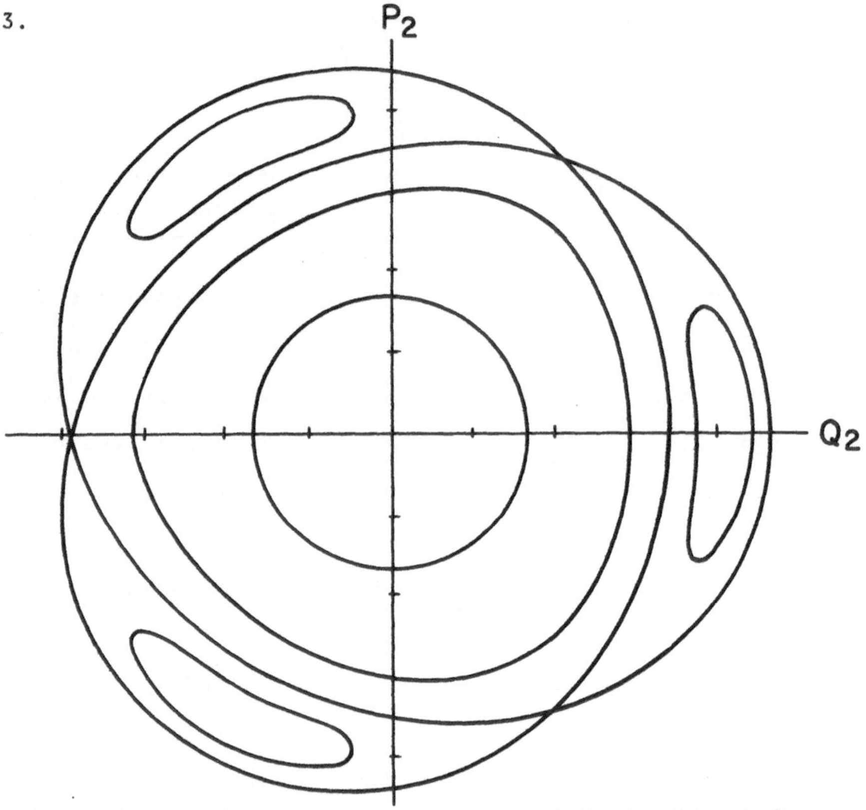

FIGURE 3. Trajectory intersections with the (Q_2, P_2) plane generated by Hamiltonian (3.3).

Note that the intersection plane is now chosen to be the (Q_2, P_2) plane. For Hamiltonian (3.3), $(3J_1 + 2J_2)$ is a second constant of the motion, and the (Q_2, P_2) plane is covered with smooth curves. As before the $\lambda = 0$ circles are for the most part only slightly distorted into the $\lambda \neq 0$ ovals. However in the resonance zone, which occurs in the region $5J_1 \cong (1+12J_2)$ where $2\Omega_1 \cong 3\Omega_2$, the tori are again grossly distorted. In a sense, the unperturbed tori in this resonant zone are bent into pretzels such that the (Q_2, P_2) plane intersection of each $\lambda \neq 0$ torus yields three distinct, crescent curves (called a chain of islands). The three invariant points at the center of each crescent region represent a single, stable periodic orbit; and the three, self-intersection points on the separatrix represent a single, unstable periodic orbit.

These two examples suffice to give the general picture. Isolated resonant interactions serve to distort the unperturbed phase space by introducing in pairs new stable and unstable periodic orbits. In addition they introduce a totally new type of torus bearing trajectories which asymptotically approach some unstable periodic orbit. Finally let us note that as the nonlinearity parameters α and λ increase or equivalently as the total energy H increases, the resonant, crescent zones move about and their widths increase. This immediately leads one to speculate on the behavior of unperturbed trajectories lying in a region simultaneously affected by two over-lapping, "isolated" resonant interaction terms.

We thus choose to investigate the Hamiltonian

$$H = J_1 + J_2 - J_1^2 - 3J_1J_2 + J_2^2 + \alpha J_1J_2\cos(2\phi_1-2\phi_2)$$
$$+ \lambda J_1J_2^{3/2} \cos(2\phi_1-3\phi_2), \qquad (3.4)$$

where H_0 is the same as in the previous two examples and where the two cos-terms previously studied in isolation now both act simultaneously. Figure 4 plots selected trajectory intersection curves generated by Hamiltonian (3.4) at the energy $E = 0.18$.

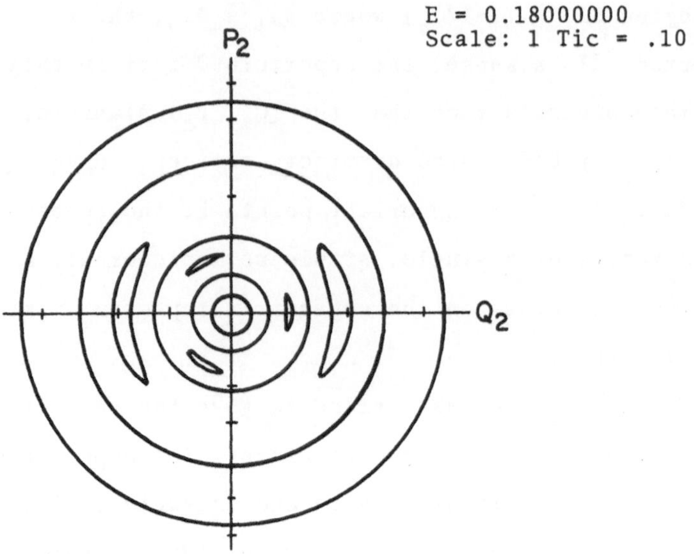

FIGURE 4. Trajectory intersection plane for Hamiltonian (3.4) at energy E = 0.18.

At this energy it appears that the $2\Omega_1 = 2\Omega_2$ and the $2\Omega_1 = 3\Omega_2$ resonant zones are widely separated, since at least one preserved torus yielding a smooth oval (centered on the origin) separates the two regions. Figure 5 shows curves in the (Q_2, P_2) plane at the energy E = 0.2, and we note, as anticipated, that the widths of the two disjoint crescent regions have increased. Indeed here the two resonance regions are very close to overlap, and the computer detects the narrow chain of five islands which is shown as well as an even narrower, nearby chain of seven islands which is not shown.

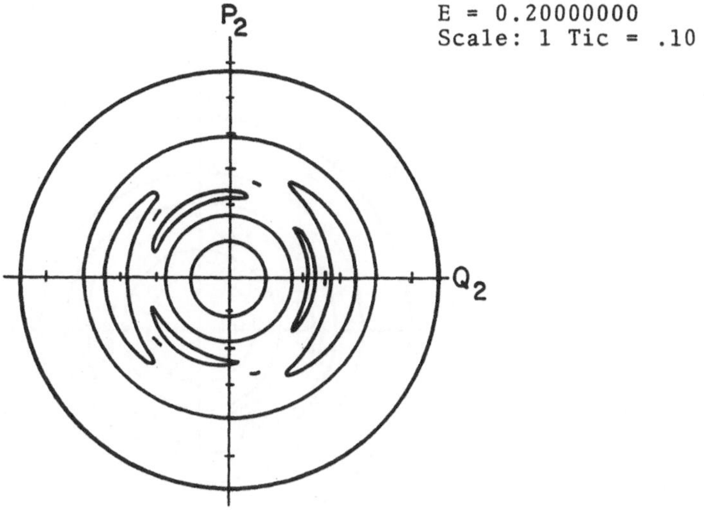

FIGURE 5. Trajectory intersection plane for Hamiltonian (3.4) at energy E = 0.20.

As the energy is slightly increased to E = 0.2095, one anticipates resonance overlap not only between the primary $2\Omega_1 = 2\Omega_2$ and $2\Omega_1 = 3\Omega_2$ zones but also among the narrower higher order resonance zones of the type detected by the computer. Trajectories in this region of overlap are thus being affected simultaneously by several resonant interaction terms and hence are being asked to respond simultaneously to the "pull" of the various stable periodic orbits and the "repulsion" of the various unstable periodic orbits. In addition those separatrix tori bearing trajectories asymptotic to the various unstable periodic orbits would (speaking somewhat naively) now intersect each other, adding to the chaos in the region. As a consequence trajectories in this region develop acute vertigo and wander aimlessly throughout the overlap region. In particular, all the isolated dots in Fig. 6 correspond to the intersection points of a single orbit.

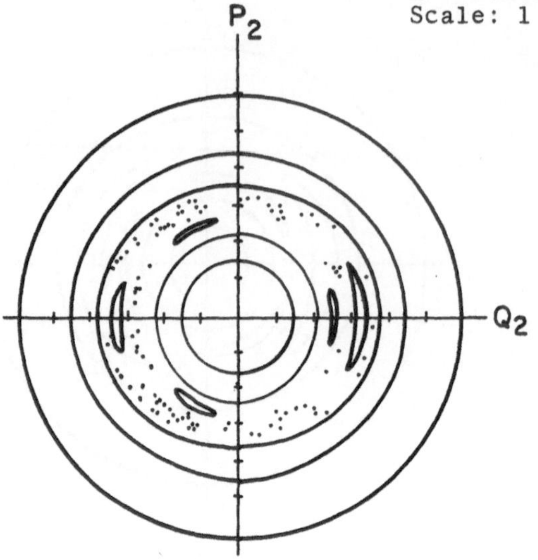

FIGURE 6. Trajectory intersection plane for Hamiltonian (3.4) at energy E = 0.2095. All the isolated dots were generated by a single trajectory.

As the energy is further increased, the macroscopic stochastic zone increases in size, eventually filling almost all the available area in the (Q_2, P_2) plane.

Henon and Heiles[7] consider the more physically realistic looking Hamiltonian

$$H = (1/2)(P_1^2 + P_2^2 + Q_1^2 + Q_2^2) + Q_1^2 Q_2 - (1/3)Q_2^3. \qquad (3.5)$$

The behavior of trajectories generated by Hamiltonian (3.5) at three distinct, increasing energies is surveyed in Fig. 7, 8, and 9. Figure 7 shows the system deep within the region of KAM stability; Fig. 8 shows the system at an energy for which the stochastic zone covers about 30% of the available area; finally Fig. 9 shows that almost all trajectories are highly stochastic when the nonlinearity becomes sufficiently large.

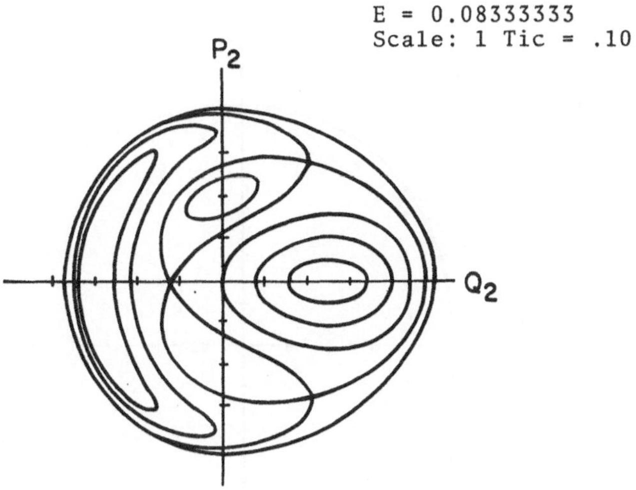

FIGURE 7. Trajectory intersection plane for the Henon-Heiles system at energy E = $^1/_{12}$. Here all curves appear to be smooth although very high computer accuracy reveals that there are narrow regions having complex structure (not shown).

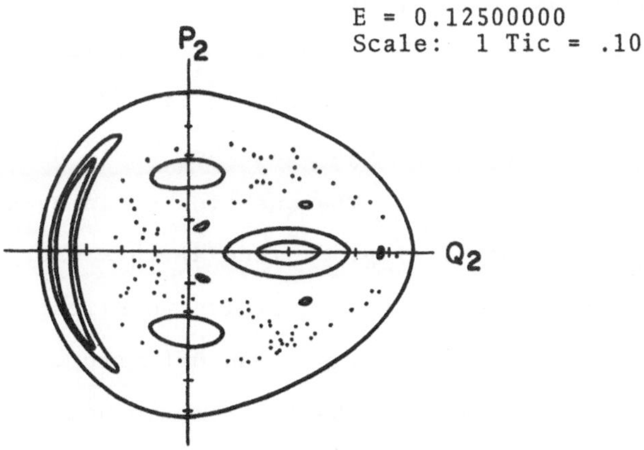

FIGURE 8. Henon-Heiles trajectory intersection plane at energy E = $^1/_8$. Here all the dots are generated by a single orbit.

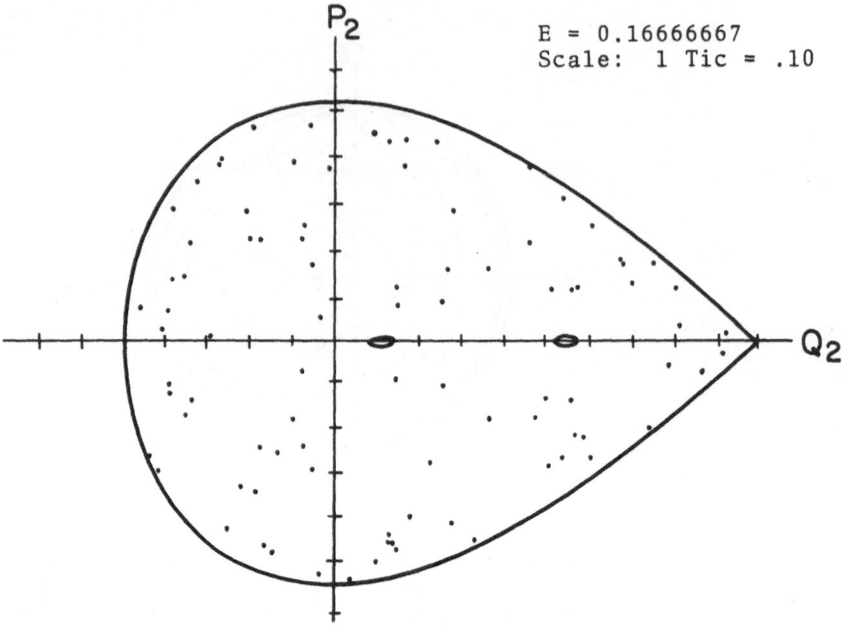

FIGURE 9. Henon-Heiles system at energy E = 1/6. Again all the dots were generated by a single trajectory.

It perhaps should be emphasized in regard to these figures that the computer has rather poor resolving power; hence Fig. 7 does not show the microscopic zones of instability implied by the KAM theory. As a consequence some of the curves in Fig. 7 might in fact be merely quasi-curves in the sense that they would disappear with improved integration accuracy. The question of integration accuracy becomes especially acute in Fig. 9 since at this energy computation of a trajectory is incredibly sensitive to slight changes in initial data as well as to accumulating round-off error.

In this section we have attempted not only to demonstrate that widespread stochasticity occurs for most oscillator system trajectories as the nonlinearity becomes large but also to present at least an intuitive understanding of this stochasticity in terms of resonance overlap. Our presentation is deeply rooted in the KAM theory. Even though in its region of validity the KAM theory precludes wholesale stochastic behavior and therefore might be thought to exclude statistical mechanics, the small set of erratic trajectories moving under the influence of many resonances forms the doorway through which statistical mechanics can be regained as the KAM assumptions are violated. Thus the KAM theorem is very likely to become one of the cornerstones in the foundations of statistical mechanics. Regardless, in this section we have established that stochastic behavior can occur in systems having a large nonlinearity; we have yet to show that just a little nonlinearity can go a long way toward producing the same effects. This we do in the next chapter.

CHAPTER IV. STOCHASTICITY FOR SMALL NONLINEARITY

In order to demonstrate that widespread stochasticity can occur in physically interesting systems even as the nonlinearity tends to zero, we must consider violation of KAM Assumption (ii). Let us again introduce the canonical transformation given by eq. (2.2) into Hamiltonian (2.1), but here let us not include all the pure J_k terms in H_0. Instead let us write Hamiltonian (2.3) in the form

$$H = H_0(J_1,\ldots,J_N) + V_3(J_1,\ldots,\phi_N) + V_4(J_1,\ldots,\phi_N) + \ldots, \quad (4.1)$$

where $H_0 = \sum_{k=1}^{N} \omega_k J_k$ with ω_k being the constant, positive frequencies of the harmonic approximation. H_0 in Hamiltonian (4.1) violates the KAM assumption (ii) with a vengeance since the $\Omega_k(=\partial H_0/\partial J_k)$ do not depend on J_k. Consequently the Jacobian of the Ω_k with respect to the J_k is very much identically equal to zero.

Because we choose to consider such a violent abrogation of KAM condition (ii), it is convenient to discuss Hamiltonian (4.1) in terms of a slightly modified version of the KAM theorem which has been stated by Arnold[5]. For Hamiltonian (4.1), Arnold rigorously proves that most trajectories lie on smooth N-dimensional integral surfaces (tori) embedded in the 2N-dimensional phase space provided, among other things,

(a) all the V_k or, equivalently, the total energy is sufficiently small, and

(b) the harmonic frequencies ω_k do not satisfy low order resonance conditions of the form

$$\sum_{k=1}^{N} n_k \omega_k = 0$$

for integers n_k such that

$$\sum_{k=1}^{N} |n_k| \le 4.$$

Speaking formally, the Arnold conditions are sufficient to insure that Hamiltonian (4.1) can be brought to a form for which the KAM theorem applies. Speaking less formally, Arnold Condition (a) is equivalent to KAM condition (i), and nothing new is to be learned from its violation. On the other hand violation of Arnold Condition (b) is an especially virulent means of violating KAM condition (ii).

In order to understand this point, suppose we try to eliminate (via a canonical transformation) the angle dependent terms in V_3 (no pure J_k terms exist in V_3), then the transformation would involve possibly zero, J_k-independent denominators of the form $(\omega_i \pm \omega_j \pm \omega_k)$ which could be zero for all values of J_k. A similar, though slightly more complicated, argument applied to V_4. In essence the validity of Arnold Condition (b) prevents the appearance of small denominators in the angle-elimination process before V_5; but by then the pure J_k terms in the lower order V_4---which cannot be eliminated---have made the frequencies of the motion depend on the J_k with (in general) a nonzero Jacobian, thus allowing us to apply the KAM theorem. We now see that violation of Arnold condition (b) is especially violent because it allows the resonant V_3 and/or V_4 terms, if present, to distort every unperturbed torus of H_0, since, up to the order considered, the frequencies on all tori have the same J_k-independent, commensurate values. Moreover V_3 and/or V_4 will in general contain many resonant terms which overlap throughout the allowed phase space. Thus we anticipate that violation of Arnold condition (b) will lead to wildly stochastic systems.

Since it is difficult in a few words to make the general situation clear, let us discuss a specific, physically interesting example. Consider the simple three-oscillator Hamiltonian

$$H = J_1 + 2J_2 + 3J_3 + \gamma[\alpha J_1 J_2^{1/2} \cos(2\phi_1 - \phi_2) + \lambda(J_1 J_2 J_3)^{1/2} \cos(\phi_1 + \phi_2 - \phi_3)], \qquad (4.2)$$

where $\omega_1 = \omega_2/2 = \omega_3/3$ and γ is an overall nonlinearity parameter introduced to "turn-off" the nonlinearity as $\gamma \to 0$. These frequencies were chosen to simulate the linear acoustic region of the dispersion curve for solids. The two cos-terms represent the everywhere overlapping, resonant, three-phonon interactions $2\omega_1 \rightleftarrows \omega_2$ and $(\omega_1+\omega_2) \rightleftarrows \omega_3$. It may be argued[9] that Hamiltonian (4.2) is the simplest model of a solid which can exhibit stochastic behavior; whatever the case, Hamiltonian (4.2) clearly violates Arnold condition (b).

If we now introduce the time dependent canonical transformation

$$J_k = J_k, \ \phi_k = \theta_k + kt, \ k = 1, 2, 3, \tag{4.3}$$

then Hamiltonian (4.2) becomes

$$H = \gamma[\alpha J_1 J_2^{1/2} \cos(2\theta_1 - \theta_2) + \lambda(J_1 J_2 J_3)^{1/2} \cos(\theta_1 + \theta_2 - \theta_3)]. \tag{4.4}$$

But γ is merely a multiplicative factor in Hamiltonian (4.4); thus γ affects only the time scale of the motion. Consequently if Hamiltonian (4.4) generates stochastic trajectories, they will persist even in the limit as $\gamma \to 0$, excluding $\gamma = 0$ of course. Finally let us note that $(J_1+2J_2+3J_3)$ is a second constant of the motion for this three-oscillator system, and hence we may reduce[9] Hamiltonian (4.4) to a Hamiltonian system having only two degrees of freedom. This then allows us to graphically present trajectory intersection curves for this system just as we did in the previous section.

Figures 10 and 11 present typical trajectory intersection curves for two distinct values of $(J_1+2J_2+3J_3)$, all other parameters being the same for both figures. The allowed area in Fig. 10 contains about 70% smooth curves, while Fig. 11 has only about 10% smooth curves. It is to be emphasized that the stochasticity in these figures is independent of the value of γ. Thus we have at last demonstrated that a little nonlinearity can indeed go a very long way.

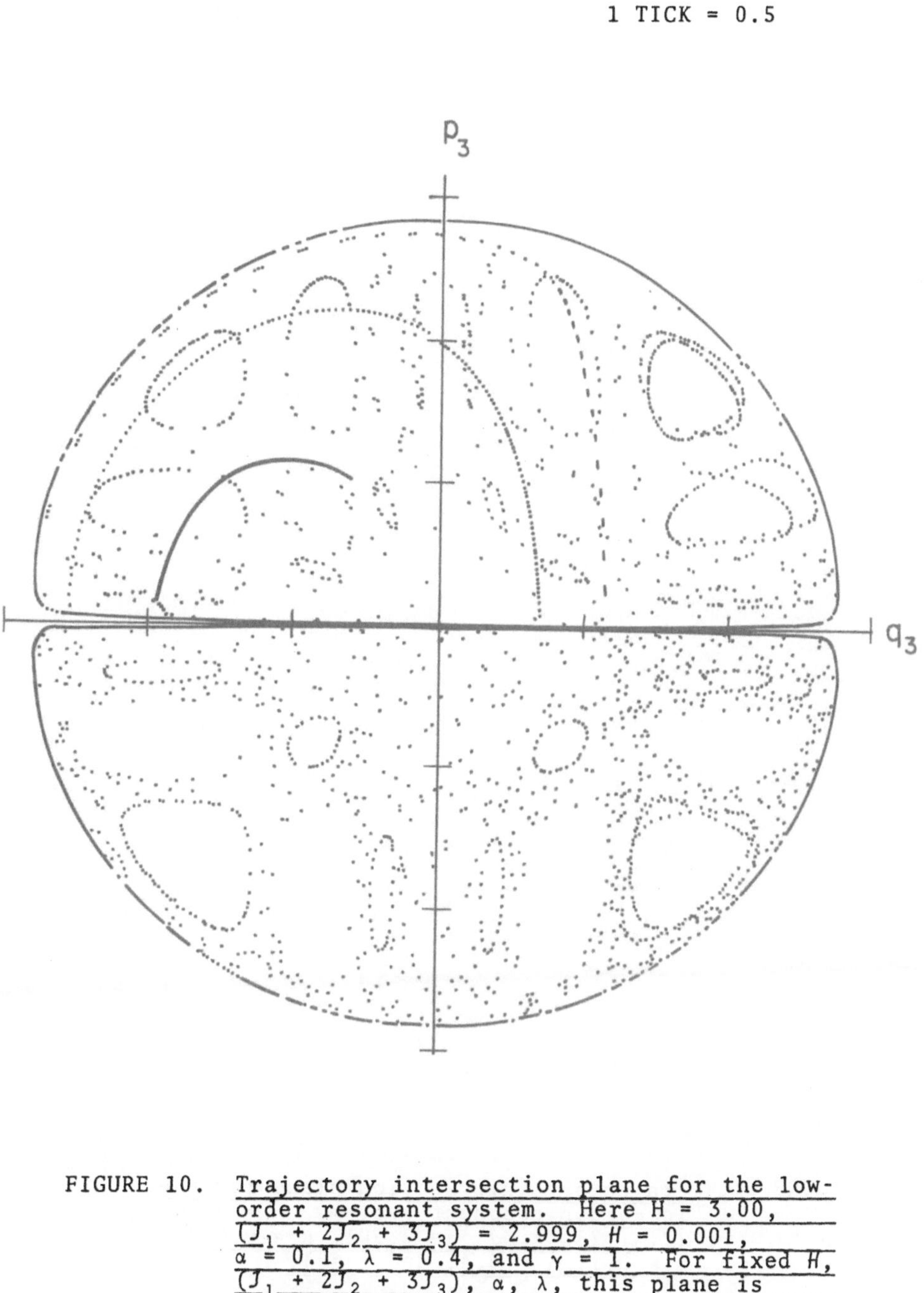

FIGURE 10. Trajectory intersection plane for the low-order resonant system. Here H = 3.00, $(J_1 + 2J_2 + 3J_3) = 2.999$, $\mathcal{H} = 0.001$, $\alpha = 0.1$, $\lambda = 0.4$, and $\gamma = 1$. For fixed $\mathcal{H}$, $(J_1 + 2J_2 + 3J_3)$, α, λ, this plane is invariant as $\gamma \to 0$.

1 TICK = 0.5

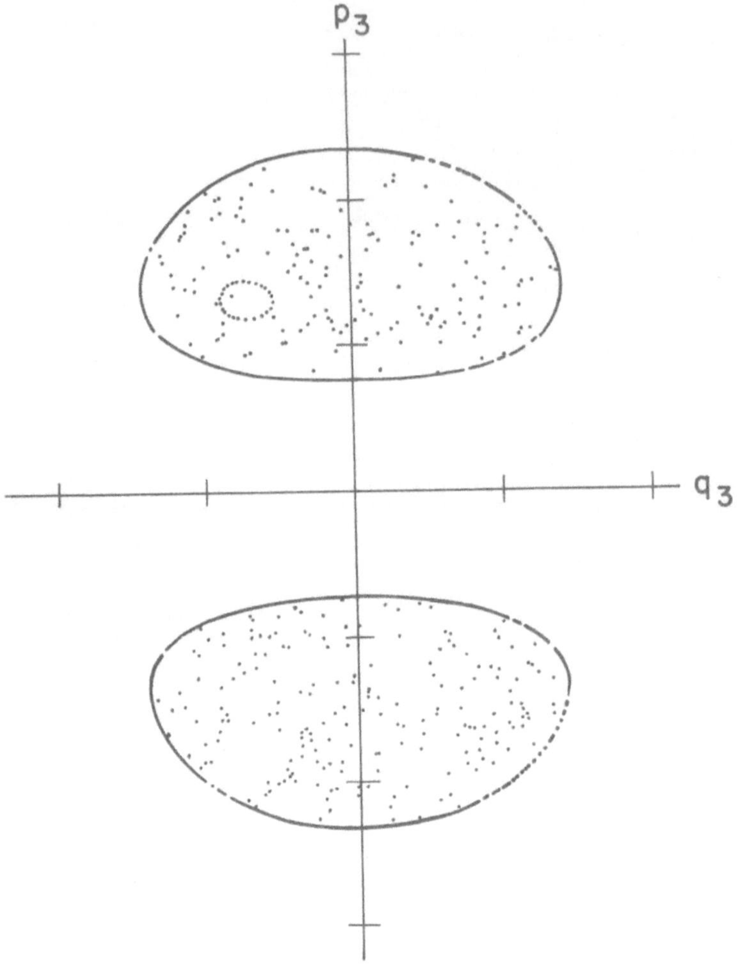

FIGURE 11. Trajectory intersection plane for the low-order resonant system. $H = 3.00$ $(J_1 + 2J_2 + 3J_3) = 2.901$, $H = 0.099$, $\alpha = 0.1$, $\lambda = 0.4$, and $\gamma = 1$.

The fact that this three-oscillator system is at best stochastic only on a four-dimensional subspace of the five-dimensional energy surface and that it is not completely stochastic even there is in actuality only a minor defect. As the number N of oscillators becomes large, the number of overlapping, resonant interaction terms goes up as N^2; hence the $N = 3$ system should exhibit the minimum stochasticity to be expected for low order resonant systems. Equally the fact that stochasticity occurs only on a (2N-2)-dimensional surface rather than a (2N-1)-dimensional surface becomes less significant as N becomes large. Finally it should be mentioned that Hamiltonian (4.2) is for physical oscillator systems the generic, rather than a specific, case, since low order resonance linking all degrees of freedom is assumed to be ubiquitous in physical oscillator systems.

CHAPTER V. EXPONENTIATING TRAJECTORIES

In the preceding sections, we have characterized stochastic and non-stochastic behavior in terms of resonance overlap or its lack, and we have pictured this difference in terms of randomly scattered points versus smooth curves generated by trajectory intersections with some specified plane. Let us now take a different and perhaps more fundamental view. Statistical mechanics argues that one of the essentials for irreversibility is that a system very quickly "forgets" its initial state (Q_i, P_i). Such "forgetting" would be expected to occur if---for some specified short time interval---the slightest change in the initial state led to a wildly different final state. In such an event, even the slightest imprecision in measuring the (Q_f, P_f) of the final state would generate complete ignorance of the initial state (Q_i, P_i). It matters not that the system itself is actually in a precise final state and therefore "remembers" its precise initial state; the crucial point is that measuring the final state (Q_f, P_f) as accurately as we can yields little or no information about the initial state. In particular and more importantly, the system would yield these same measured (Q_f, P_f) values---hence the same thermodynamic quantities---had it started in any of a widespread class of initial states. In a very real sense, the many-body problem is assumed to be so pathological that only God could cope with precise classical mechanics; mere man must be contented with statistical mechanics. Fortunately this assumed pathology can be empirically demonstrated and studied on a computer using even simple nonlinear oscillator systems in their region of KAM instability where the system trajectories are wildly sensitive to even the slightest variation of initial conditions.

We may vividly illustrate this latter point by plotting the phase space separation distance between two initially close trajectories as a function of time. In Fig. 12 we plot separation distance versus time for four orbit-pairs initially started about 10^{-7} units apart in a region of KAM stability.

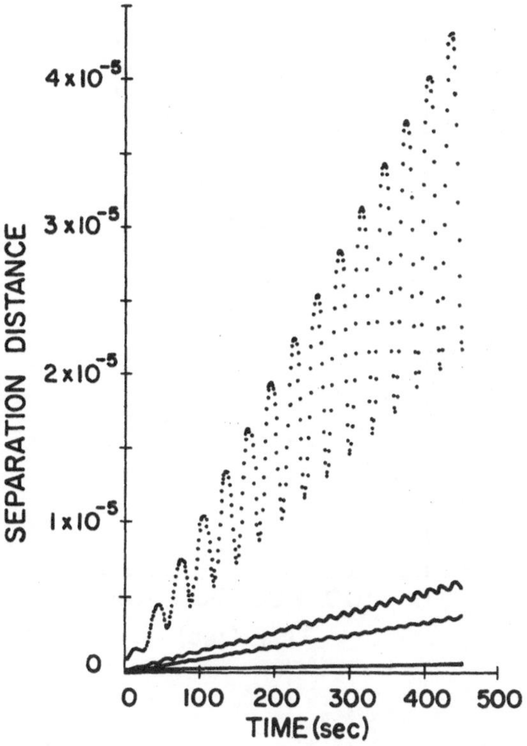

FIGURE 12. A plot of the phase space separation distance between two initially close orbits as a function of time for a typical two-oscillator system. The curves for four orbit pairs are plotted. All orbit-pairs are initiated in a smooth curve region of the trajectory intersection plane.

Here we notice a linear growth of separation distance with time. This is the type of "streaming" separation distance growth we would expect for two initially close runners moving at slightly different speeds. In Fig. 13 using a log-plot, the two generally upper-lying curves show the "exponential" growth with time for two orbit-pairs started in a region of KAM instability.

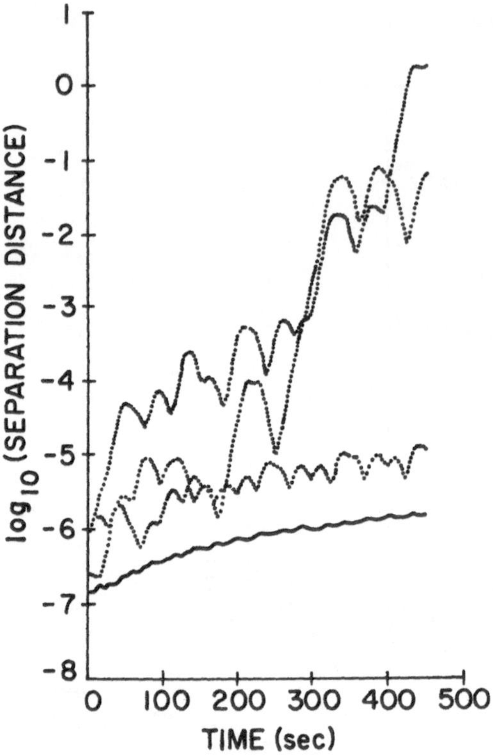

FIGURE 13. A log-plot of the separation distance versus time for four distance orbit-pairs. The two generally upper lying curves show the typical "exponential" separation for stochastic orbit-pairs. The two generally lower lying curves are for orbit-pairs initiated in a non-stochastic region.

For comparison the two lower-lying curves are for two orbit-pairs in a region of KAM stability. The time scale should be ignored; the essential point is that the unstable orbit-pairs increase their separation distance by a factor of 10^6 during the same time that the stable orbit-pairs increase by only about a factor of 10. Thus in regions of KAM instability, nonlinear oscillator systems appear to "exponentially forget" their initial state.

Exponentiating trajectories, here observed for oscillator systems, lie at the heart of Sinai's proof[10] of the ergodic and mixing character of the hard sphere gas. Miller[11] has produced empirical (computer) evidence showing that star clusters have exponentiating trajectories, and D. V. Anosov and Ja. Sinai[12] prove that a wide class of physical systems have this property. The interested reader might also wish to consult the papers by Alekseev[13], the excellent review article by Wightman[14], and the review text by Arnold and Avez[15]. The point to be emphasized here is that exponentiating trajectories in phase space may be the fundamental property of physical systems approaching equilibrium.

CHAPTER VI. AREA PRESERVING MAPPINGS

We have now demonstrated that systems having two degrees of freedom exhibit much of the complexity of the many-body problem and that we may graphically survey this complexity through plots of trajectory intersections with specified planes. Let us now observe that these graphs in fact merely represent area-preserving mappings of a plane onto itself which are generated by solving differential equations[15]. In this view the earlier investigations were merely studies on the generic properties of area-preserving mappings. But if we wish to investigate the generic properties of mappings, it would be much easier to study directly algebraic mappings of the form $x_1 = x_1(x,y)$, $y_1 = y_1(x,y)$. This not only eases the problem of numerical computation, it also allows us to use various mapping theorems to further illuminate the pathology of the many-body problem.

Let us begin by considering the area-preserving mapping T given by

$$x_1 = x\cos\alpha - y\sin\alpha \tag{6.1a}$$

$$y_1 = x\sin\alpha + y\sin\alpha \tag{6.1b}$$

The origin in the (x,y) plane is an elliptic fixed point of T, and the invariant curves, corresponding to the earlier trajectory intersection curves, of T are circles. T maps each point (x,y) through the same angle α (called the rotation number) on its invariant circle $(x^2 + y^2)^{1/2}$. Let us now rewrite T as

$$r_1 = r \tag{6.2a}$$

$$\theta_1 = \theta + \alpha \tag{6.2b}$$

Then generalize slightly and introduce a twist into the rotation via

$$r_1 = r \tag{6.3a}$$

$$\theta_1 = \theta + \alpha(r). \tag{6.3b}$$

Circles are still invariant curves of this twist mapping T_1, but now the rotation angle α depends on $r = (x^2 + y^2)^{1/2}$. Both T and T_1 generate mappings corresponding to "integrable" differential equation systems which have smooth trajectory intersection curves everywhere.

The central question now concerns whether or not the invariant curves of T_1 persist under perturbations. Thus let us consider the mapping T_2 given by

$$r_1 = r + f(r,\theta) \qquad (6.4a)$$

$$\theta_1 = \theta + \alpha(r) + g(r,\theta), \qquad (6.4b)$$

where f and g are periodic 2π in θ. For this mapping Moser[16] has shown that provided---speaking loosely---f and g are sufficiently small then those invariant curves for which

$$\left|\frac{\alpha(r)}{2\pi} - \frac{p}{q}\right| \geq \frac{\varepsilon}{q^{5/2}} \qquad (6.5)$$

persist under the perturbations f and g, being only slightly distorted. In analogy with the tori of differential equation systems, those unperturbed curves having rational rotation numbers are destroyed by the perturbation.

But here we know a little more about the stochastic regions. Let us consider the unperturbed circles bearing rational rotation numbers. Such circles are made up of the invariant points of $(T_2)^n$, i.e. T_2 applied n times, where $\alpha(r) = 2\pi(m/n)$. Now when f and g are nonzero but small, Birkhoff[17] has shown that the complete circle of fixed points of $(T_2)^n$ does not persist. Indeed only 2n of the fixed points persist with half being elliptic and half being hyperbolic. Thus T_2 will in general yield a mapping of the type pictured in Fig. 14 taken from the paper by Arnold.[5] Let us note that both the Moser and the Birkhoff theorems will be valid in some neighborhood of each elliptic fixed point of Fig. 14.

FIGURE 14. Typical graph of the mapping T_2 of eq. (6.4), showing only the gross features. The waving, self-intersecting curves are discussed in Chapter VII.

Consequently this whole picture repeats itself on successive microscopic levels, "boxes" within "boxes" ad infinitum! One thus has the incredible result that even simple quadratic, algebraic mappings (see eq. (6.6) below) or the mappings generated by two-oscillator systems such as Hamiltonian (3.5) can yield the wonderous complexity of Fig. 14. Using the highest computer accuracy, for example, one can see the beginning levels of this complexity in Fig. 7. Rather than show an improved version of Fig. 7 however, let us consider the mapping T_3 given by

$$x_1 = x\cos\alpha - y\sin\alpha + x^2\sin\alpha \tag{6.6a}$$

$$y_1 = x\sin\alpha + y\cos\alpha - x^2\cos\alpha \tag{6.6b}$$

which has already been studied by Henon[18]. In Fig. 15 we graph T_3 using $\cos\alpha = 0.24$ and notice that at this level of computer accuracy the curves around the elliptic points appear smooth.

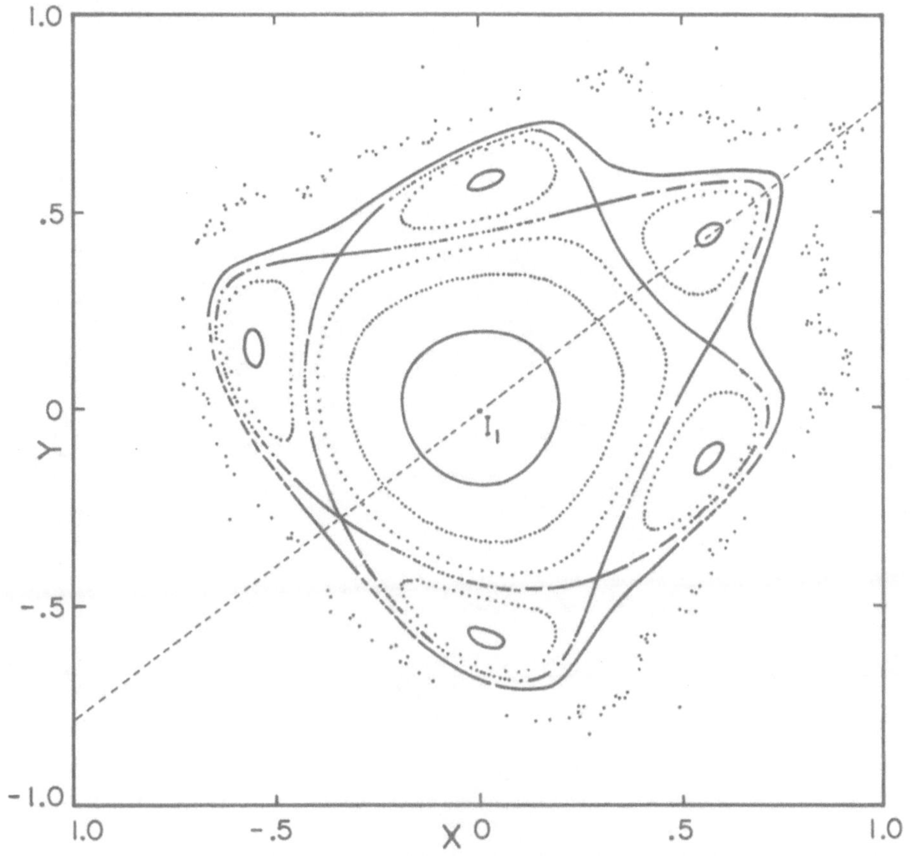

FIGURE 15. The Henon mapping T_3 for $\cos\alpha = 0.24$. Here the invariant curves look smooth.

However in Fig. 16, which is a magnified look at the small region near the right-most hyperbolic fixed point, we see some of the underlying complexity.

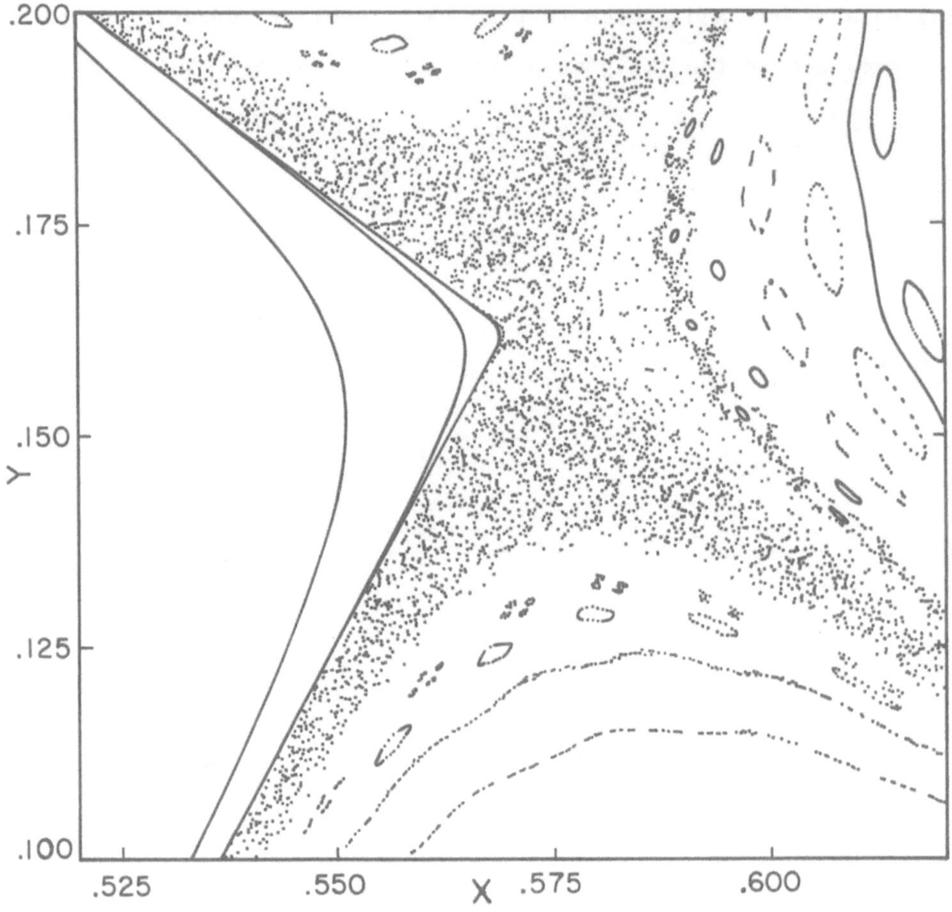

FIGURE 16. A magnified look at the right-most hyperbolic fixed point in Fig. 15 showing the very complex structure which exists on a microscopic scale.

CHAPTER VII. PATHOLOGY OF AREA-PRESERVING MAPPINGS

In order to begin to understand the source of some of this pathology, let us consider an "integrable" area-preserving mapping T_4 (not explicitly written out here) which yields exact, invariant curves everywhere, of which a few are shown in Fig. 17.

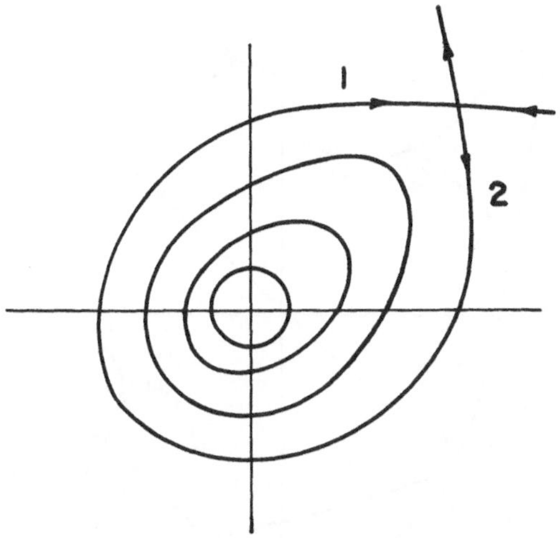

FIGURE 17. Invariant curves for an "integrable" mapping T_4 which generates smooth invariant curves everywhere. Attention should be focused on the self-intersecting separatrix curve whose self-intersection point is an ordinary hyperbolic fixed point.

Let us ignore the elliptic fixed point at the origin and focus our attention on the invariant curve passing through the hyperbolic fixed point. On part 1 of this curve, successive points generated by the mapping move toward the fixed point; while on part 2, they move away from the fixed point. Moreoever we may generate part 1 by calculating backward iterates $(T_4)^{-n}$ of some point on the curve near the fixed point; while part 2 may be generated by calculating forward iterates $(T_4)^n$ of some point on the curve near the fixed point. In particular, we observe in Fig. 17 that the forward and backward curve segments thus generated must join smoothly into a single curve. However if we now "turn on" the "nonintegrable" perturbations f and g, we find, as shown[19] in Fig. 18, that the forward and backward curves now intersect each other.

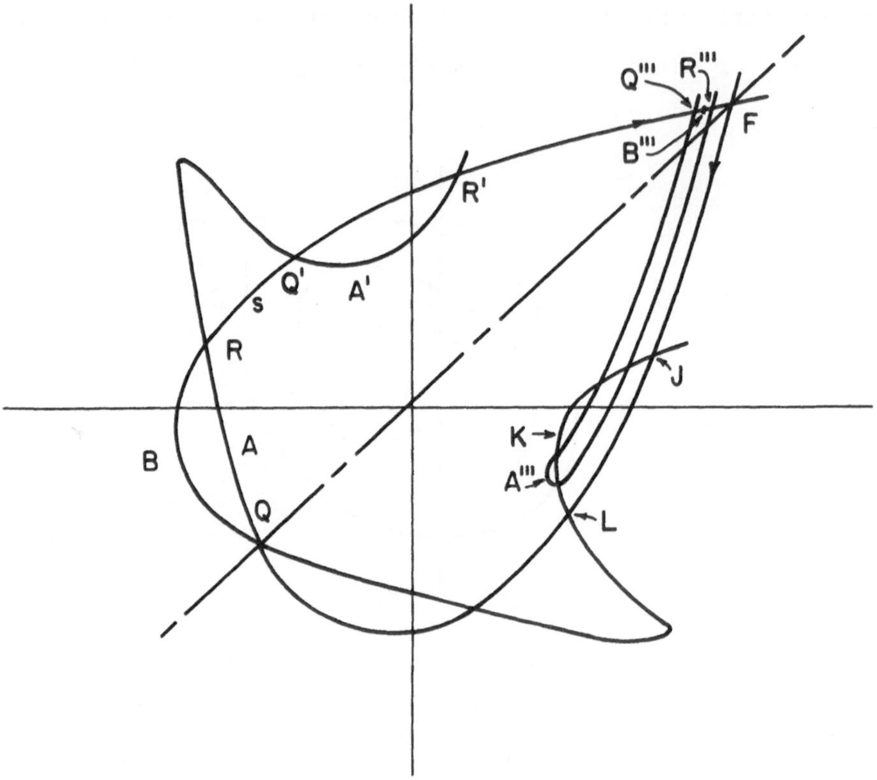

FIGURE 18. Invariant curves for a small "non-integrable perturbation" of the mapping T_4 in Fig. 17. Note that the oval part of the separatrix curve in Fig. 17 has "split" into two intersecting curves.

Moreover each of these curves begins to oscillate wildly as they asymptotically approach the fixed point. The segment Q'''A'''R''' in Fig. 18, for example, lies on the forward interated curve while the segment JKL lies on the backward iterated curve.

The nature of the self-intersecting curves of Fig. 14, which were ignored earlier, can perhaps now be made clear. Here the backward iterated curve coming from one hyperbolic fixed point intersects the forward iterated curve from an adjacent hyperbolic fixed point; and in Fig. 14 we have an example of the typical break-up of the "integrable" system separatrix that appeared in Fig. 3. Several other features of Fig. 18 are noteworthy. The area QARB eventually maps into thinner and thinner "area-filaments" of which Q'''A'''R'''B''' is an early example. Thus points initially close together quickly map apart. As the nonlinear parameter increases, these forward and backward "separatrix" curves intersect at an ever increasing angle, "plowing-up" increasing portions of the stable region around the elliptic fixed point. Finally when there are many overlapping resonances in a region, the oscillating "separatrix" curves belonging to one of the resonances not only intersect each other, they also intersect the "separatrix" curves belonging to the other resonances. Thus it is quite understandable that the set of iterates of a given point in such a region appear to occur at random.

Even the elliptic fixed points can turn hyperbolic when the nonlinearity is sufficiently strong, either because the nonlinearity itself is large or because there are overlapping, low-order resonances. The hyperbolic fixed point---called hyperbolic (with reflection)---arising from a converted elliptic fixed point is more pathological than an ordinary hyperbolic fixed point. In Fig. 17 for example on the ingoing separatrix curves, iterates of a point on one side of the fixed point asymptotically approach the fixed point from the same side. On the other hand were this a hyperbolic (with reflection) fixed point, then iterates of a point on the ingoing separatrix would still asymptotically approach the fixed point, but now succeeding iterates would lie on alternate sides of the fixed point. As an example consider the two adjacent hyperbolic (with reflection) fixed points shown[20] in Fig. 19. Here the point labeled A_{n+8} is the next iterate of the point labeled A_n, and we note that A_n lies on one ingoing "separatrix" curve while A_{n+8} lies on the other ingoing "separatrix".

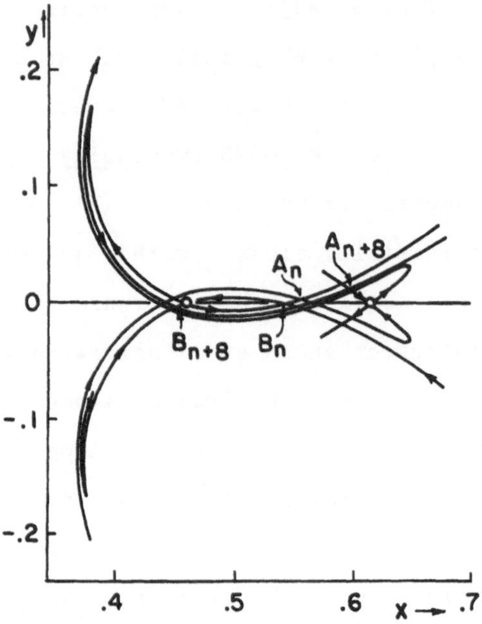

FIGURE 19. An example of hyperbolic (with reflection) fixed points, showing an adjacent pair of such points.

It should also be noted that the "separatrix" curves in Fig. 19 are much wilder than their cousins in Fig. 18.

The two hyperbolic (with reflection) points shown in Fig. 19 are members of a family of sixteen fixed points. Eight are hyperbolic (with reflection) and eight are ordinary hyperbolic. If the nonlinearity parameter of the mapping were lowered, then this family would become a chain of eight islands with eight elliptic fixed points separated by eight ordinary hyperbolic fixed points. This allows the possibility that in the highly nonlinear stochastic regions of a mapping, all (or almost all) fixed points are hyperbolic. Without going into the details, we mention that J. M. Greene[21] has used a computer to generate strong empirical evidence that this is indeed the case. In a region containing only a large number of hyperbolic

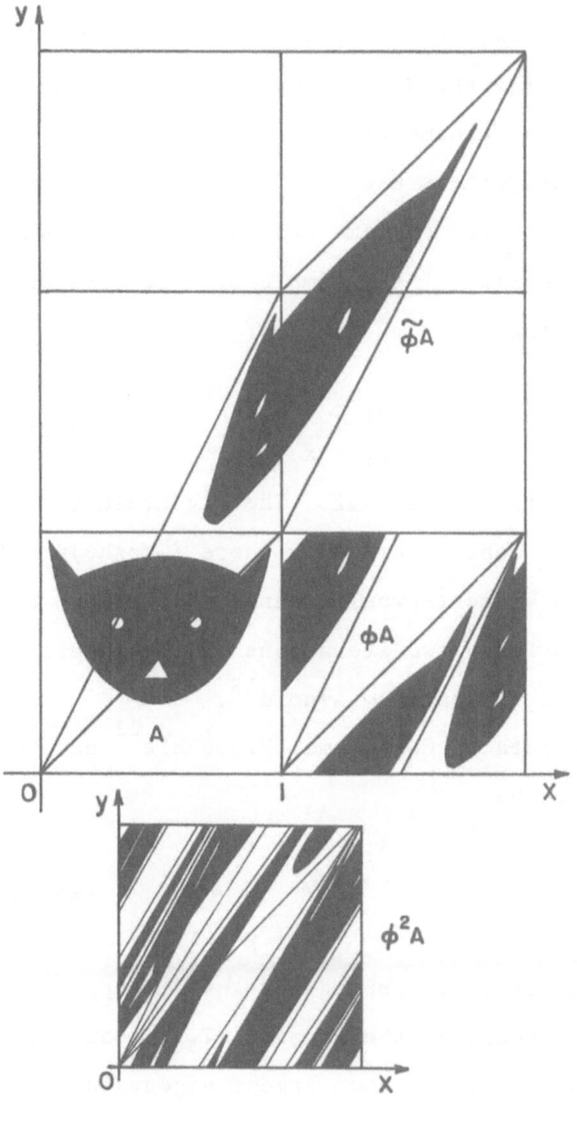

FIGURE 20. An example of a mapping which is mixing. After only two iterations of the mapping, one already has a crazy, mixed-up cat.

fixed points, the iterates of almost every point would be expected to appear quite random, and two points started close together would be expected to diverge apart at an "exponential" rate. Thus we return once again to the fundamental property - exponentiating orbits.

Arnold[15] and others use this exponentiating property as the defining condition for systems called C-systems, which may be shown to be ergodic and mixing. As a simple example consider the area-preserving mapping

$$x_1 = x + y \qquad\qquad (6.7a)$$
$$\qquad\qquad (\mathrm{mod}\ 1)$$
$$y_1 = x + 2y \qquad\qquad (6.7b)$$

of the unit square onto itself. The eigenvalues of the matrix involved in this mapping are $e^{\pm\theta}$, where $(2\cosh\theta) = 3$. Thus we expect that small area elements would change their shape exponentially. The mixing caused by only two iterations of the mapping in eq. (6.7) is shown in Fig. 20, taken from Arnold[15].

Using the idea of C-systems, Froeschle[22] has investigated the mapping T_5 given by

$$x_1 = x + \lambda\sin y \qquad\qquad (6.8a)$$
$$\qquad\qquad (\mathrm{mod}\ 2\pi)$$
$$y_1 = x + y + \lambda\sin y. \qquad\qquad (6.8b)$$

A typical graph of this mapping for $\lambda = 1.3$ appears in Fig. 21. Froeschle now linearizes the mapping $(T_5)^n$ about some arbitrary point (x,y) and then calculates the largest eigenvalue λ_n of the linearized $(T_5)^n$ for each n. If the arbitrary point (x,y) lies in an "exponentiating" stochastic region then we would expect

$$\theta_n = (\ell n\lambda_n)/n \qquad\qquad (6.9)$$

to approach a constant value as $n \to \infty$.

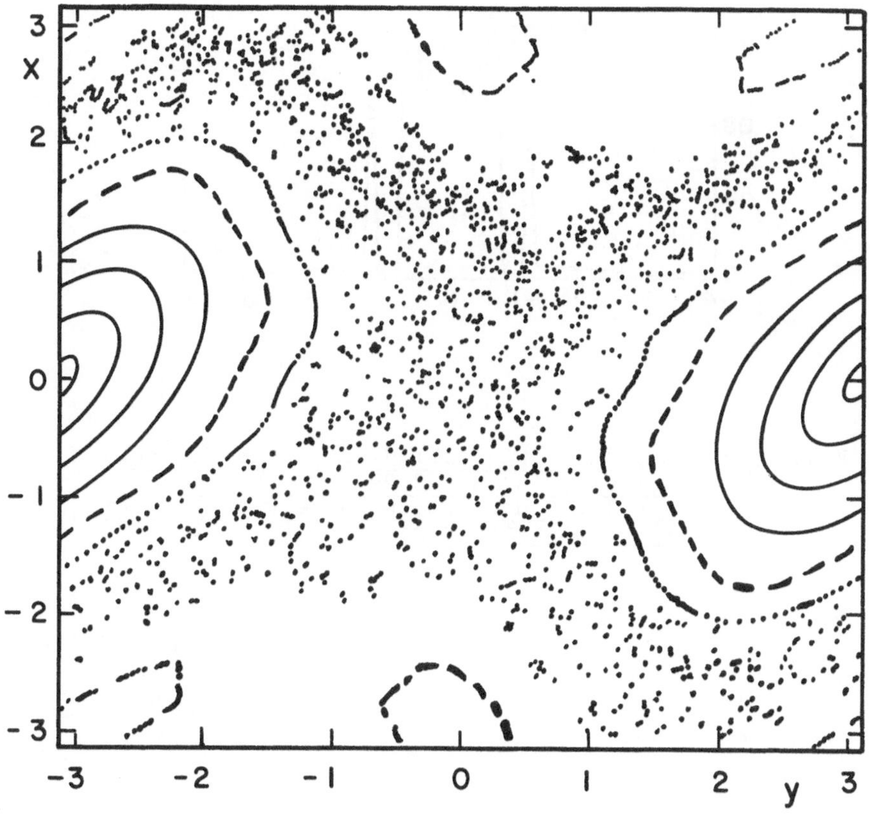

FIGURE 21. A typical graph of the mapping considered by Froeschle.

If the point (x,y) lies in a non-stochastic curve region, then one would expect θ_n to approach zero as $n \to \infty$. In order to minimize fluctuations, Froeschle calculates the average θ_n given by

$$\mu_n \equiv \frac{1}{n} \sum_{k=1}^{n} [(\ell n\lambda_k)/k]. \tag{6.10}$$

In Fig. 22 we reproduce Froeschle's graph of μ_n versus y for the mapping T_5, where n = 2,000, x = 0, and $\lambda = 1.3$. One notices that the peaked region of μ_n in Fig. 22 occurs just as would be predicted from Fig. 21. Froeschle has thus obtained empirical computer evidence that this mapping behaves like a C-system in the stochastic regions.

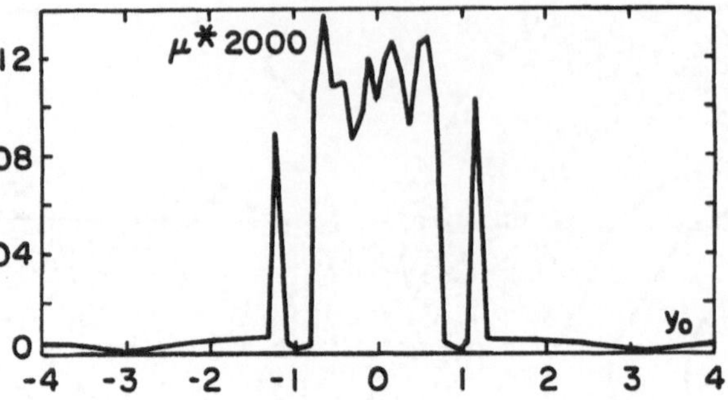

FIGURE 22. A plot of μ_n (called μ_n^* in the figure) versus y for n = 2,000, x = 0, and λ = 1.3. Here μ_n is large when y is in the stochastic regions of Fig. 21.

CHAPTER VIII. CONCLUSIONS

In this paper we have attempted to illustrate, using very simple examples, that stochastic behavior expected to be characteristic of all many-body problems which exhibit an approach to equilibrium. We have thereby tried to demonstrate as simply as possible the significance for statistical mechanics of the new mathematical results which have been obtained during the past decade. Perhaps the most fundamental characteristic of these systems is the "exponential" stirring of phase space of the type envisioned by Gibbs as leading to irreversible behavior. It remains to be seen whether or not these new insights are sufficient to lead to new solutions to all the old problems in statistical mechanics.

References:

[1] A. N. KOLMOGOROV: Dokl. Akad. Nauk. SSSR 98, 527 (1954).

[2] V. I. ARNOLD: Russian Math. Surveys 18, 9 (1963).

[3] J. MOSER: Nachr. Akad. Wiss. Gotingen, II Math Physik Kl. 1 (1962).

[4] E. MERZBACKER: Quantum Mechanics (John Wiley and Sons, Inc. New York, 1964).

[5] V. I. ARNOLD: Russian Math. Surveys 18, 85 (1963).

[6] For the casual reader, the words "stochastic" and "erratic" may be taken to be synonyms. The technical minded reader should refer to the series of articles by V. M. ALEKSEEV listed in Reference 13.

[7] M. HENON AND C. HEILES: Astron. J. 69, 73 (1964).

[8] G. WALKER AND J. FORD: Phys. Rev. 188, 416 (1969).

[9] J. FORD AND G. H. LUNSFORD: Phys. Rev. A1, 59 (1970).

[10] Ja. SINAI: in Statistical Mechanics, Proceedings of the IUPAP Meeting, Copenhagen, 1966, edited by T. A. Bak (W. A. Benjamin Inc., New York, 1967), p. 559.

[11] R. H. MILLER: Astrophys. J. 140, 250 (1964).

[12] D. V. ANOSOV AND Ja. SINAI: Russian Math. Surveys 22, 103 (1967).

[13] V. M. ALEKSEEV: Math. USSR Sbornik 5, 73 (1968): 6, 505 (1968): 7, 1 (1969).

[14] A. S. WIGHTMAN: in Statistical Mechanics at the Turn of the Decade, edited by E. G. D. COHEN (Marcel Dekker, New York, 1971) pp. 1-32.

[15] V. I. ARNOLD AND A. AVEZ: Ergodic Problems of Classical Mechanics (W. A. Benjamin Inc., New York, 1968). Also see J. M. GREENE: J. Math. Phys. 9, 760 (1968).

[16] J. MOSER: Mem. Am. Math. Society 81, 1 (1968).

[17] G. D. BIRKHOFF: Dynamical Systems (American Mathematical Society Publications, New York, 1927).

[18] M. HENON: Quart. Appl. Math. 27, 291 (1969).

[19] Figure 18 is reproduced from a letter to the author by L. JACKSON LASLETT of the Lawrence Radiation Laboratory, Berkeley.

[20] L. JACKSON LASLETT: AEC Research and Development Report NYO-1480-101, New York University, 1968 (unpublished).

[21] J. M. GREENE: J. Math. Phys. 9, 760 (1968).

[22] C. FROESCHLE: Astron. and Astrophys. 9, 15 (1970).

NONEQUILIBRIUM THERMODYNAMICS, DISSIPATIVE STRUCTURES, AND BIOLOGICAL ORDER *

Jack S. Turner
Center for Statistical Mechanics and Thermodynamics
The University of Texas at Austin

FORWARD

The study of chemical instabilities, especially as possible mechanisms for the appearance of spatial, temporal, and functional order in biology, is experiencing a period of tremendous growth at the present time. Consequently any general review of the subject is quite likely to be at least partly out of date by the time it appears. Moreover, workers in the field necessarily draw ideas from such diverse classical disciplines as chemical kinetics, thermodynamics, hydrodynamics, mathematical stability theory, and various areas of biology, plus quite recent developments in nonlinear irreversible thermodynamics. Depending on one's level of sophistication in each of these areas, therefore, following the original literature in this field may prove to be a difficult task, particularly for the newcomer. For these reasons this report will not constitute a survey of current endeavors in the field. Rather, in a presentation of the general theoretical foundations for a unified study of dissipative instabilities, the treatment of all essential topics has been made as self-contained as possible. In particular relevant aspects of thermodynamics, mathematics, and biology are presented in sufficient detail that reference elsewhere should not be essential to understanding. In elaborating the general theory, a few simple examples are presented in detail to indicate the wide variety of ordered behavior which chemical systems may exhibit beyond instability. Finally, to illustrate the nature of applications to living systems, two important classes of phenomena are discussed both in a biological context and in terms of specific theoretical models which exhibit many features of the observed behavior. In this way, it is hoped that the many excellent reviews of the proliferating literature in this exciting field may be made accessible to as wide an audience as possible.

---

*This article is based in part on a series of lectures presented by G. Nicolis at the Advanced School for Statistical Mechanics and Thermodynamics, Austin, Texas, 1970.

INTRODUCTION [1]

From a physico-chemical point of view perhaps the most striking feature of biological systems is the high level of complexity and organization exhibited by living things. To maintain a state of such high coherence, even in the simplest organism, requires an enormous number of metabolic and synthetic reactions together with complex mechanisms for controlling the rate and timing of various life processes. It is generally accepted that the present biological order reflects the effect of organization acquired through long (prebiological and biological) evolution: living systems have evolved from "simple" to extremely organized and complex forms. At the same time such essential processes as metabolism, synthesis, regulation, and the storage and transfer of information imply a highly heterogeneous distribution of matter within living cells due to chemical reactions and active transport. At first sight both (evolutionary and spatial) characteristics of living things appear to be at variance with the fundamental laws of macroscopic physics, in particular the second law of thermodynamics.

According to "classical" (equilibrium) theory the natural evolution of a thermodynamic system yields ultimately an equilibrium state of maximum disorder. For isolated systems the second law ensures a monotonic increase of entropy to a maximum at the equilibrium state, precluding the formation of any ordered (i.e., low entropy) structure. For nonisolated systems, exchanging energy and/or matter with the surroundings, the situation is different. At sufficiently low temperatures, only low-lying system energy states will be populated, and the possibility then exists for formation of ordered states of low entropy. This ordering principle accounts for the appearance of *equilibrium structures* such as crystals. It cannot, however, explain the formation of biological structures: The probability that at ordinary temperatures a macroscopic number of molecules is assembled spontaneously to produce the highly complex structures and coordinated functions which characterize living systems is vanishingly small. Hence the apparent contradiction between biological order and the laws of physics cannot be resolved by the classical methods of equilibrium statistical mechanics and thermodynamics.

This failure of equilibrium theory suggests that the answer must lie in an extension of the concept of order to nonequilibrium situations. Indeed, the very existence in biological systems of nonequilibrium constraints, from the solar energy gradient at the macroscopic extreme to gradients of concentration or chemical potential on the cellular level, implies such a possibility. Since living organisms exist under open system conditions, the overall change in entropy of a biological system includes a flow term $\Delta_e S$ due to exchange with the surroundings and a production term $\Delta_i S$ due to irreversible processes in the system. Therefore it is possible for such a system to reach a low-entropy state and to maintain it indefinitely provided it can attain a steady state (i.e., $\Delta S = 0$) such that $d_e S = -d_i S \leq 0$. In principle, then, a negative entropy flow can maintain a low-entropy state, which by definition must be a steady nonequilibrium state. A particularly simple example of such nonequilibrium order is provided by a thermal diffusion experiment, in which a temperature gradient imposed on an initially homogeneous mixture produces and maintains a separation of matter corresponding to a decrease of entropy from the initial state.

Such qualitative arguments obviously do not "prove" the compatibility of biological evolution and order with physical law. Nevertheless they do indicate that a new "ordering principle" appropriate to nonequilibrium situations may indeed provide the answer. In addition to demonstrating that living systems satisfy (globally) the second law of thermodynamics, however, one would like also to understand how such states of low entropy and high coherence may be formed and how they are maintained. In recent years these questions have attracted a great deal of attention. Due to an important extension of nonequilibrium thermodynamics by Glansdorff and Prigogine [2], it is now possible to approach these problems systematically in terms of thermodynamic variables of direct experimental interest. A principal conclusion of the new theory is that there exists a class of systems which may exhibit two completely different types of behavior: a tendency toward maximum disorder in some circumstances, and in others the appearance of coherent behavior. Near thermodynamic equilibrium destruction of order always yields a unique steady state which is stable with respect to all possible disturbances. Such a state lies on the "thermodynamic branch" of solutions to the corresponding macroscopic kinetic equations, and represents the continuous extension into the nonequilibrium domain of the equilibrium state. Far from equilibrium, however, creation of order may occur spontaneously in open systems, obeying specific nonlinear kinetic laws, which are maintained beyond the domain of stability of the thermodynamic branch. The traditional theories of equilibrium and linear irreversible thermodynamics treat the first type of behavior. Within the framework of generalized nonlinear thermodynamics [2] has emerged a nonequilibrium concept of order through fluctuations [1] in terms of which the second type of behavior, appropriate in particular to biological order and evolution, may be understood.

Nonequilibrium transitions (to ordered regimes) occurring beyond instabilities have long been known in hydrodynamics (e.g., laminar to turbulent fluid flow). In purely dissipative systems, however, one would at first sight expect the "usual" thermodynamic tendency toward maximum disorder to be the rule. On the other hand, in examples of the latter type, especially chemically reacting mixtures under open system conditions, nonlinearities may arise in a practically unlimited variety of ways. In particular biological mechanisms involving auto- or cross-catalysis, induction, repression, etc., present an enormous wealth of possibilities for interesting (i.e., ordered) nonlinear behavior. Thus it is not now surprising that models of a wide variety of purely dissipative systems are found to exhibit many of the same features which characterize hydrodynamic instabilities [1,2]. Such systems (like their hydrodynamic counterparts) evolve beyond instability to qualitatively new types of states, called dissipative structures [2], which are created and maintained by the dissipative, entropy-producing processes occurring inside the system.

In this review the object will be to examine the concept of structure formation through nonequilibrium instabilities in the context of selected aspects of the problem of biological order. To this end the subject divides naturally into two distinct pieces. In Part One the necessary thermodynamic background is obtained by reviewing the "classical" theory of irreversible thermodynamics and the recent generalization to nonlinear situations. A few simple mathematical models are then analyzed in order to provide an idea of the types of behavior which can be expected from nonlinear chemical networks operating far from equilibrium. For insight into the mechanism of instabilities and transitions we then consider the behavior of fluctuations around nonequilibrium steady states and the related question of deterministic vs stochastic description for systems exhibiting macroscopic instabilities. In PART TWO are presented a number of biological problems to which the considerations of PART ONE can be applied. Specifically we consider particular mathematical models for two much-studied types of biological organization at the cellular level: the regulation of metabolic and of biosynthetic processes. Following detailed discussion of these two examples, we conclude by listing a number of recent reviews, to which the reader is referred for additional applications and citations to the original literature.

PART ONE

THERMODYNAMICS OF IRREVERSIBLE PROCESSES: THE ORIGIN AND STABILITY OF NONEQUILIBRIUM ORDER

INTRODUCTION *

To introduce the type of system which we discuss in our study of irreversible phenomena, consider first an isolated system (i.e., a system of constant energy and number of particles). It is easy to see that the description of this system reduces to a problem of classical thermodynamics. Indeed, the second law of thermodynamics implies here that the entropy of the system is non-decreasing: $(dS)_{E,N} \geq 0$. The system evolves therefore to a uniquely determined time-independent state which is necessarily the state of thermodynamic equilibrium. It is characterized by the property that the entropy S (a thermodynamic potential) is a maximum, which insures at the same time the stability of the equilibrium state.** We see then that for long times isolated systems attain equilibrium. If we deal therefore with "aged" systems, such as those encountered in most experimental situations, isolated systems are unlikely to exhibit interesting irreversible behavior for time scales of approach to equilibrium which are not extremely slow.

Consider now a closed system which can exchange energy with its surroundings, or an open system which can exchange both energy and matter. We may imagine an open system in contact with a number of reservoirs, each characterized by a temperature T_n, pressure p_n, and chemical potentials μ_n^α for species α. In this case, provided the

---

*Much of the content of PART ONE appears in Ref. [2]. With few exceptions, therefore, references to original literature will be made here only for more recent publications.

**This statement must be qualified in the vicinity of an instability point such as a phase transition or critical point.

external reservoirs are sufficiently large to remain time-independent, the system will also reach, for long times, a stationary regime. Here, however, two types of situations may arise: (1) The system is at thermal, mechanical, and chemical equilibrium with the reservoirs (i.e., T, p, μ^{α} are everywhere constant). The final state will then be the state of thermodynamic equilibrium, again, as for an isolated system, characterized by the extremum of a thermodynamic potential, one of the free energies F, G, etc. As before, the existence of such an extremum will guarantee the stability of the equilibrium state. (2) The system is not at thermal, mechanical, or chemical equilibrium. For example, one might impose the constraints ΔT_{12}, Δp_{12}, $\Delta\mu^{\alpha}_{12}$, etc. Provided that the constraints remain constant, the system will then reach a time-independent state which is independent of the initial preparation of the system but which is _not_ the state of thermodynamic equilibrium. Such a state we call a _steady nonequilibrium state_.

It is easy to see that such states are far from exceptional: for example, all transport processes (e.g., thermal conduction) occur as a result of the deviation of the state of the system from equilibrium, the extent of deviation being determined by the largeness of the constraints (e.g., temperature gradient) imposed on the system. Irreversibility is therefore an experimental fact. What then are the properties of these nonequilibrium states, and what is the evolution of the system near such states, especially as the constraints become large? Due primarily to recent advances in _nonlinear_ thermodynamics, _all_ such questions may now be approached within the framework of a general formulation of macroscopic thermodynamics [2].

CHAPTER I: NONEQUILIBRIUM THERMODYNAMICS. THE LINEAR THEORY.

A. Local Description of Matter.

Before we can approach this complicated problem it will be necessary to adopt a certain number of principles, or rather, a certain philosophy, underlying the description. Our main working assumption will be that the system we are interested in may be treated as a continuous medium. This is a very natural assumption which is justified for almost all practical purposes. A fluid (e.g., a dense gas or a liquid) is certainly a continuous medium as far as the evolution of macroscopic properties such as p, T, etc., is concerned. It is only in the limit of extremely low density that this assumption may break down. Such situations are therefore excluded from our discussion. In general, we say that dense systems may be treated in the continuous approximation. In particular, therefore, biological systems (e.g., a cell) are continuous media.

Having made it clear that we consider systems for which the macroscopic (observable) quantities are continuous functions of space and time, we formulate a macroscopic theory as follows: We subdivide our system into a large number of volume elements or cells, each of which must (1) contain enough molecules that its macroscopic properties can be measured or calculated by taking meaningful averages and (2) at the same time be sufficiently small that the values so obtained do not vary appreciably over the cell. Thus, we say that the cells must be microscopically large and macroscopically small. Finally, we assign to each cell appropriate values for the various macroscopic quantities of interest. From a <u>global</u> point of view, we have then specified for macroscopic quantities of interest the appropriate volume densities describing the <u>local</u> behavior of the system.

Each cell is itself an open system in which the local densities may vary in time. We begin, therefore, with a discussion of the balance equations for mass, momentum, and energy which, as we shall see, play a prominent role in thermodynamics.

B. Balance Equations. The First Law of Thermodynamics.

Consider an arbitrary volume element within the system. We are interested in the time evolution of extensive quantities I(t). We distinguish two "mechanisms" for changing I(t):

$$\frac{dI(t)}{dt} = P[I] + \Phi[I] \tag{1.1}$$

or, symbolically,

$$dI = d_i I + d_e I \ , \tag{1.2}$$

where in general only dI is a total differential of the independent variables. The terms on the right of these equations represent, respectively, the <u>production</u> of the quantity I within the volume V and the rate of <u>flow</u> of I into the volume through the boundary surface Ω. In terms of the corresponding intensive quantities (volume densities), defined by the relations

$$I(t) = \int_V f(\underset{\sim}{x},t)dV, \quad P[I] = \int_V \sigma(\underset{\sim}{x},t)dV, \quad \Phi[I] = -\int_\Omega \underset{\sim}{j}(\underset{\sim}{x},t)\cdot d\underset{\sim}{\Omega} \ , \tag{1.3}$$

the balance equation (1.1) takes the <u>local</u> form

$$\frac{\partial f(\underset{\sim}{x},t)}{\partial t} = \sigma(\underset{\sim}{x},t) - \text{div}\ \underset{\sim}{j}(\underset{\sim}{x},t). \tag{1.4}$$

As already indicated, the volume densities are assumed to be continuous functions of space and time. For simplicity this dependence will not be included explicitly in our notation.

Before going on to the specific balance equations which we shall need, we make two remarks about the last equation. The flow term will, in general, consist of separate contributions due to convection and conduction: $\underset{\sim}{j} = \underset{\sim}{j}_{conv} + \underset{\sim}{j}_{cond} = f\underset{\sim}{v} + \underset{\sim}{j}_{cond}$. The convection current corresponds to the local flow of matter (with velocity $\underset{\sim}{v}$). The conduction current occurs even in a system at rest.

Finally, we see that the balance equation (1.4) takes a particularly simple form for quantities I' which are conserved:

$$\sigma[I'] = 0 \ , \ \frac{\partial f'}{\partial t} + \text{div}\ \underset{\sim}{j}[I'] = 0 \ . \qquad (1.5)$$

Thus the conservation laws for mass, momentum, and energy (no external forces) are expressed by the equalities $\sigma[M] = 0$, $\sigma[\underset{\sim}{Q}] = 0$, $\sigma[U] = 0$. Entropy, on the other hand, is not conserved, and satisfies instead the inequality $\sigma[S] \geq 0$.

Now we recall briefly the balance equations for mass, momentum, and energy. Consider a multicomponent fluid of chemical species among which a number of reactions are possible. Each component j is characterized by a (macroscopic) local mass flow velocity $\underset{\sim}{v}_j$, mass fraction $c_j \equiv m_j/m$ $(m = \sum_j m_j)$, and density $\rho_j \equiv \rho c_j$ $(\rho = \sum_j \rho_j)$. The local center of mass velocity is given by $\rho\underset{\sim}{v} \equiv \sum_j \rho_j \underset{\sim}{v}_j$, and the diffusion velocity, $\underset{\sim}{\Delta}_j$, of component j by the decomposition $\underset{\sim}{v}_j = \underset{\sim}{v} + \underset{\sim}{\Delta}_j$. Multiplying by ρ_j gives directly the mass current density as the sum of a convection current $\rho_j \underset{\sim}{v}$ and a conduction (diffusion) current $\rho_j \underset{\sim}{\Delta}_j$ $(\sum_j \rho_j \underset{\sim}{\Delta}_j = 0)$.

If we label the chemical reactions by the index r, then we have a system of stoichiometric equations

$$0 = \sum_j \nu_{jr} M_j \ , \ r = 1,2,\ldots \qquad (1.6)$$

expressing conservation of mass for individual reactions. Here, M_j is the molecular weight of species j and ν_{jr} is the stoichiometric coefficient for component j in reaction r (defined positive for products, negative for reactants). Finally we define w_r to be the rate of chemical reaction r and write the mass balance equation for component j:

$$\frac{\partial \rho_j}{\partial t} = \sum_r \nu_{jr} M_j w_r - \text{div}(\rho_j \underset{\sim}{v} + \rho_j \underset{\sim}{\Delta}_j) \ . \qquad (1.7)$$

It is most important to note here that the reaction rate w_r is in general a <u>nonlinear</u> function of the densities (concentrations) so that the mass balance equations for different species are nonlinearly coupled through the rates.

If we now sum over species in Eq. (1.7), we see that the source term vanishes (conservation of mass) according to relations (1.6). The diffusion current vanishes also and we obtain the familiar continuity equation for matter,

$$\frac{\partial\rho}{\partial t} + \mathrm{div}\, \rho\underset{\sim}{v} = 0 \ . \tag{1.8}$$

According to classical continuum mechanics the equation of motion for a continuous medium is written

$$\rho\frac{d\underset{\sim}{v}}{dt} = \rho\underset{\sim}{F} - \mathrm{div}\, \underset{\approx}{P} \ . \tag{1.9}$$

Here $\underset{\approx}{P}$ is the pressure tensor of the medium and $\underset{\sim}{F}$ the force per unit mass due to external fields. If we introduce the so-called hydrodynamic derivative or Stokes' operator

$$\frac{d}{dt} = \frac{\partial}{\partial t} + \underset{\sim}{v} \cdot \mathrm{grad} \ , \tag{1.10}$$

then the equation of motion takes the standard form (1.4) of a balance equation

$$\frac{\partial}{\partial t} (\rho\underset{\sim}{v}) = \rho\underset{\sim}{F} - \mathrm{div}\, (\rho\underset{\sim}{v}\underset{\sim}{v} + \underset{\approx}{P}) \ . \tag{1.11}$$

We see from this equation that the source term is

$$\sigma[\underset{\sim}{Q}] = \rho\underset{\sim}{F} \equiv \sum_j \rho_j \underset{\sim}{F}_j \ , \tag{1.12}$$

where the last equality defines the total specific force $\underset{\sim}{F}$ when the (specific) force is different for different species (e.g., electrostatic forces). The flow term contains the momentum current due to convection, $\rho\underset{\sim}{v}\underset{\sim}{v}$, and the conduction current $\underset{\approx}{P}$ corresponding to the total flow of momentum relative to the local center of mass motion. One can decompose the pressure tensor into an elastic part $\underset{\approx}{P}_e$ plus an inelastic or dissipative part $\underset{\approx}{p}$. For a fluid the elastic contribution is simply the hydrostatic pressure p, so that

$$P_{ij} = p\delta_{ij} + p_{ij} \ , \ \delta_{ij} \equiv \begin{cases} 1, & i=j \\ 0, & i\neq j \end{cases} \ . \tag{1.13}$$

Clearly, the inelastic contribution must vanish at equilibrium. As is customary in hydrodynamics, we assume that the pressure tensor is symmetric.

We have noted that the mass balance equation may be linear when the reaction rates are linear in the densities. Because of the presence of the convection current, however, the momentum balance equation is <u>always</u> nonlinear whenever there is local (macroscopic) mass motion. This fact is particularly important for all problems relating to hydrodynamics.

Now we turn to the balance equation for energy. If we multiply the local equation of motion (1.11) by $\underset{\sim}{v}$, we obtain directly a balance equation for the local center of mass kinetic energy:

$$\frac{\partial}{\partial t}(\tfrac{1}{2}\rho v^2) = \rho \underset{\sim}{v} \cdot \underset{\sim}{F} + \underset{\approx}{P}:\text{grad } \underset{\sim}{v} - \text{div}(\tfrac{1}{2}\rho v^2 \underset{\sim}{v} + \underset{\approx}{P} \cdot \underset{\sim}{v}) \ . \quad (1.14)$$

For the potential energy of component j due to the external forces, the production term corresponds to the power per unit volume developed against the forces $\underset{\sim}{F}_j$ along the motion $\underset{\sim}{v}_j$. If ω_j represents the potential energy per unit mass of component j, then the appropriate balance equation takes the form*

$$\frac{\partial}{\partial t} (\rho_j \omega_j) = -\rho_j \underset{\sim}{v}_j \cdot \underset{\sim}{F}_j - \text{div} (\rho_j \omega_j \underset{\sim}{v}_j) \ . \quad (1.15)$$

Combining source terms for center of mass kinetic energy and potential energy due to external forces, and taking into account the definition of the diffusion velocity, we have

$$\sigma[E_k + E_p] = \sum_j \rho_j \underset{\sim}{\Delta}_j \cdot \underset{\sim}{F}_j + \underset{\approx}{P}:\text{grad } \underset{\sim}{v} \ . \quad (1.16)$$

From a macroscopic point of view we therefore define a quantity E, the internal energy, by the requirement for total energy conservation

$$\sigma[U] \equiv \sigma[E_k + E_p + E] = 0 \ . \quad (1.17)$$

By definition, the conduction contribution to the flow of internal energy is the heat flow $\underset{\sim}{W}$. If e is the specific internal energy, then

---

*Here we have assumed that potential energy is a property of matter and therefore moves by convection. For energy exchange by radiation, for example, we must add a conduction term, the Poynting vector of the electromagnetic field.

the appropriate balance equation is

$$\frac{\partial}{\partial t}(\rho e) = \sum_j \rho_j \Delta_j \cdot \underset{\sim}{F}_j - \underset{\approx}{P}:\text{grad } \underset{\sim}{v} - \text{div }(\rho e \underset{\sim}{v} + \underset{\sim}{W}) \; . \qquad (1.18)$$

It is easy to verify that this equation is simply the first law of thermodynamics in local form. Indeed, for a simple fluid in an external field such that $\underset{\sim}{F}_j = \underset{\sim}{F}$ is the same for all components, then the first production term vanishes and we have, at thermal equilibrium,

$$\frac{de}{dt} = -\frac{1}{\rho}\,\text{div } \underset{\sim}{W} - p\frac{dv}{dt} \; . \qquad (1.19)$$

Here $v = 1/\rho$ is the specific volume and we have used Eqs. (1.10), (1.13), and the fact that the dissipative part of the pressure tensor vanishes at equilibrium. Finally, at equilibrium the pressure and density are uniform, so that integrating over the entire volume yields the standard expression for the first law of thermodynamics $dE = dQ - pdV$.

C. Local Equilibrium. The Second Law of Thermodynamics.

Having now the necessary mechanical background we are prepared to set up our thermodynamic theory. The starting point is of course the second law of thermodynamics, which states that the entropy of an isolated system is nondecreasing. If we write the total entropy differential for a system as a balance equation [see Eq. (1.2)], then the statement of the second law appropriate to closed or open systems (which provide the situations of interest to us here) takes the form

$$d_iS = dS - d_eS \geq 0 \; . \qquad (1.20)$$

Here d_iS is the entropy production due to irreversible processes within the system, d_eS is the entropy flow due to interactions with the outside world, and the equality holds for reversible processes.

Before incorporating the thermodynamic criterion for irreversibility into our local description, we require a general statement which specifies for arbitrary nonequilibrium situations how such thermodynamic quantities as entropy (also T, p, μ_j, etc.) are related to the relevant thermodynamic independent variables (e.g., E, V, c_j). We take a given thermodynamic system subdivided into cells as before. At thermodynamic equilibrium the quantities T, p, μ_j, and S are well-defined in terms of E, V, c_j. Away from equilibrium, however, it is necessary to redefine these quantities. We assume, therefore, that the thermodynamic dependent variables defined locally for each subsystem of a globally nonequilibrium system depend on the appropriate local thermodynamic independent variables exactly as at equilibrium. In particular, the local entropy will have the same functional dependence on the local variables, e, v, c_j as at equilibrium. Thus, if s is the specific entropy, the familiar Gibbs relation for $s(e, v, c_j)$ takes the local form

$$Tds = de + pdv - \sum_j \mu_j dc_j \ . \tag{1.21}$$

The local equilibrium assumption and its explicit representation (1.21) provide the basis for the local formulation of irreversible thermodynamics [3]. The question of its microscopic justification and interpretation has been treated by Prigogine, who showed that the assumption of local equilibrium implies the dominance of dissipative processes over purely mechanical processes. Specifically, the (local) molecular distribution functions $f_1(\underset{\sim}{x}, \underset{\sim}{v}; t)$ may deviate only slightly from their equilibrium forms. The microscopic studies of Prigogine and coworkers [4] show, for systems in which f_1 is sufficiently close to local equilibrium, that the Boltzmann entropy definition $\rho s = -k\int d\underset{\sim}{v} f_1 \ln f_1$ implies the local Gibbs relation (1.21). Thus, the local theory of irreversible thermodynamics is indeed applicable to such systems. Just how large is the class of such systems? We shall simply indicate the answer by mentioning a few systems which may exhibit typical far-from-equilibrium behavior. For quite complicated systems of chemical reactions governed by highly nonlinear rate equations and having extremely large affinities the local theory is adequate provided the reactions are not too fast. Similarly, all effects described by the Navier-Stokes equations, including in particular hydrodynamic instabilities, are within the domain of validity of the local equilibrium theory.

D. Entropy Balance. Local Formulation of the Second Law.

Using the local equilibrium assumption we may now proceed to introduce the second law of thermodynamics into the local theory. Following the procedure of Sec. I-B, we rewrite the entropy balance equation (1.20) in the form

$$dS = P[S] + \Phi[S] \quad , \quad P[S] \geq 0 \ . \tag{1.22}$$

Introducing now the entropy production density (local entropy production) $\sigma[S]$ according to Eq. (1.3), we have

$$P[S] \equiv \frac{d_iS}{dt} = \int dV\sigma[S] \quad , \quad \sigma[S] \geq 0 \ , \tag{1.23}$$

where the inequality expresses the second law in local form.

What we now require is a connection between $\sigma[S]$ and the appropriate local thermodynamic variables. Such a connection is established by substituting into the local Gibbs relation (1.21) expressions for de, dv, dc_i from the balance equations obtained in Sec. I-B. To get Eq. (1.21) into the required form, we introduce the time derivative explicitly, multiply by ρ, and use the hydrodynamic derivative (1.10) to transform

$$T\,\frac{ds}{dt} = \frac{de}{dt} + p\,\frac{dv}{dt} - \sum_j \mu_j \frac{dc_j}{dt} \tag{1.24}$$

into

$$T\frac{\partial(\rho s)}{\partial t} + T\mathrm{div}(\rho \underset{\sim}{v} s) = \frac{\partial(\rho e)}{\partial t} + \mathrm{div}(\rho \underset{\sim}{v} e) + p\mathrm{div}\,\underset{\sim}{v} - \sum_j \mu_j\left[\frac{\partial\rho_j}{\partial t} + \mathrm{div}(\rho_j \underset{\sim}{v})\right] \ . \tag{1.25}$$

Substituting on the right hand side from balance equations (1.7) and (1.18) and rearranging a few of the terms gives a local entropy balance equation of the form

$$\frac{\partial}{\partial t}(\rho s) = \sigma - \mathrm{div}\,\underset{\sim}{j}_s \ , \tag{1.26}$$

where the (local) entropy flow is

$$\underset{\sim}{j}_s = \rho s \underset{\sim}{v} + \frac{\underset{\sim}{W}}{T} - \frac{1}{T}\sum_j \rho_j \underset{\sim}{\Delta}_j \mu_j \tag{1.27}$$

and the (local) entropy production is

$$\sigma = \underset{\sim}{W} \cdot \text{grad}\,\frac{1}{T} - \sum_j \rho_j \underset{\sim}{\Delta}_j \cdot [\text{grad}(\frac{\mu_j}{T}) - \frac{\underset{\sim}{F}_j}{T}] - \frac{1}{T}\,\underset{\approx}{p}:\text{grad}\,\underset{\sim}{v} + \frac{1}{T}\sum_r w_r A_r \ . \qquad (1.28)$$

In writing the last equation we have defined the affinity of chemical reaction r in terms of the chemical potentials (per unit mass) of the species involved in that reaction:

$$A_r \equiv -\sum_j \nu_{jr}\mu_j M_j \ . \qquad (1.29)$$

The decomposition of the local entropy change into flow and production terms has been performed here so that the entropy production vanishes at thermodynamic equilibrium. This decomposition, however, is by no means unique, as alternative definitions of the heat flow, for example, are sometimes convenient [3]. The entropy flow (1.27) consists of a convection term $\rho s\underset{\sim}{v}$ plus two conduction terms related to heat and mass transport through diffusion. The local entropy production, on the other hand, has a remarkable form which we shall exploit further in developing the thermodynamic theory.

The mathematical structure of the local entropy production (1.28) is that of a bilinear form. That is, each contribution to $\sigma[S]$ is a product of two factors, one a flow or rate of an irreversible process, and the other, the driving force, or "generalized" thermodynamic force, corresponding to that irreversible process. If we denote by J_α and X_α the flow and force, respectively, appropriate to irreversible process α, then the local entropy production takes the form

$$\sigma[S] = \sum_\alpha J_\alpha X_\alpha \geq 0 \ . \qquad (1.30)$$

The separation of the contributions in Eq. (1.28) into forces and flows is summarized in Table 1. Although other definitions of the forces and flows are possible, the entropy production itself must of course remain invariant with respect to any such new definitions.

Irreversible Process	Flow	Force	Tensor Character
Heat Conduction	heat flow $\underset{\sim}{W}$	grad $(\frac{1}{T})$	vector
Diffusion	mass flow $\rho_j \underset{\sim}{\Delta}_j$	$-\text{grad}\ (\frac{\mu_j}{T}) + \frac{\underset{\sim}{F}_j}{T}$	vector
Viscous flow	pressure tensor (momentum flow) $\underset{\approx}{p}$	$\frac{1}{T}$ grad $\underset{\sim}{v}$	second rank tensor
Chemical reaction	reaction rate w_r	affinity A_r/T	scalar

TABLE 1.

E. Closure. Linear Phenomenological Laws. Onsager's Relations.

The thermodynamic equations presented in preceding sections cannot present a closed description of a system subject to well-defined boundary conditions as long as the flows J_α appear as parameters unrelated to the forces X_α . It is essential, therefore, to supplement these general equations with specific phenomenological laws relating flows to forces.

At thermodynamic equilibrium there is no macroscopic transport of mass, momentum, or energy. At the same time the conditions of thermal equilibrium imply the absence of constraints such as temperature or concentration gradients. This means that both the generalized forces and the corresponding fluxes vanish at thermodynamic equilibrium. A natural assumption, therefore, is that near equilibrium the fluxes will depend linearly on the forces to a good approximation. The appropriate phenomenological laws, which then serve to define the domain of linear irreversible processes, take the form

$$J_\alpha = \sum_\beta L_{\alpha\beta} X_\beta , \qquad (1.31)$$

where the sum is over (coupled) irreversible processes.*

*According to the Curie-Prigogine Principle, flows and forces of different tensorial character are not coupled in an isotropic medium. For anisotropic systems, however, more general types of coupling may occur (e.g., the coupling between chemical reactions and active transport across biological membranes) [3c].

The phenomenological coefficients $L_{\alpha\beta}$ are in general functions of the thermodynamic state variables T, p, etc., and must satisfy certain conditions: $\sum_{\alpha,\beta} L_{\alpha\beta}\, X_\alpha\, X_\beta$ positive semidefinite, $L_{\alpha\beta} = L_{\beta\alpha}$. The first property follows directly from the second law. The reciprocal relation for the off-diagonal coefficients is a fundamental result due to Onsager [5], who showed that it is always possible to choose flows and forces so that the coefficient matrix is symmetric.

F. Steady States Near Equilibrium. Minimum Entropy Production.

The adoption of linear laws along with the above conditions on the $L_{\alpha\beta}$ has proven an effective method for study of many irreversible phenomena in the linear regime. Each new application, however, requires an appropriate selection of specific phenomenological laws. Hence, one is tempted to ask whether there exists any general principle, other than the second law, which can characterize nonequilibrium states without regard to the details of specific irreversible processes occurring within a system. In answer to this question has evolved an alternative approach to formulating a local theory of nonequilibrium thermodynamics: the search for variational principles and thermodynamic potentials [6].

Consider now systems which are near thermal equilibrium and which in addition are at mechanical equilibrium. If linear phenomenological laws (1.31) are assumed to be valid, then the local entropy production (1.30) becomes a quadratic form in the generalized forces

$$\sigma = \sum_{\alpha,\beta} L_{\alpha\beta} X_\alpha X_\beta \geq 0 \ . \qquad (1.32)$$

Before discussing the properties of the entropy production in the neighborhood of a nonequilibrium steady state, we must first elucidate the relationship between the generalized forces and flows and the external constraints which are responsible for maintaining such a state. Clearly if no constraints are imposed the system will ultimately reach the state of thermodynamic equilibrium, at which all forces and flows vanish identically. Whenever constraints exist, however, some forces will remain nonvanishing, as will also their conjugate flows. The effect on the thermodynamic analysis of the presence of constraints is twofold: First, not all the generalized forces will remain independent; in general several may be related through the constraints. As a consequence it may be convenient to define a new set of independent forces, and a corresponding new set of conjugate flows.* Second, the thermodynamic properties of a system subject to external constraints cannot in general be inferred from strictly local considerations, since the constraints, which may appear as boundary conditions, must be taken explicitly into account. As a rule one must take a global view, therefore, considering instead of σ[S] the integrated entropy production P[S] defined according to Eq. (1.23).

Suppose now that appropriate independent forces and flows have been selected which are compatible with the imposed constraints. Of the forces {X} a set {X"} will be fixed as long as the constraints are maintained constant, while the remaining forces {X'} will be unrestricted. The system will then attain a stationary state when all flows conjugate to unrestricted forces vanish. At the same time flows conjugate to the fixed forces take on constant values. If linear phenomenological laws hold, then $J'_{\alpha} = \sum_{\alpha} L_{\alpha\beta} X_{\beta} = 0$ at the steady state (the sum includes fixed as well as unrestricted forces). If in addition the $L_{\alpha\beta}$ are constants (to good approximation) satisfying Onsager's reciprocal relations, then the following conditions hold at the steady state:

$$\left(\frac{\partial\sigma}{\partial X'_{\alpha}}\right)_{X'_{\beta\neq\alpha}} = 2\sum_{\beta} L_{\alpha\beta} X_{\beta} = 0 \quad \text{and} \quad \left(\frac{\partial^2\sigma}{\partial X'^2_{\alpha}}\right)_{X'_{\beta\neq\alpha}} = 2L_{\alpha\alpha} \geq 0. \quad (1.33)$$

These relations are symbolic in that the global entropy production, and a more general variational problem, are often involved [3a]. If the conditions outlined in this section are met, however, the conclusion is unchanged: The entropy production, considered as a function of the gen-

*The selection here of appropriate forces and flows is analogous to the choice of an optimum coordinate system for problems in analytical mechanics.

eralized forces, is minimum at the steady state attained by a system subject to time-independent constraints. To illustrate this fundamental theorem, several simple examples are discussed in Appendix B.

G. Stability of Steady States. Impossibility of Ordered Behavior Near Equilibrium.

Such general statements as the theorem of minimum entropy production (above) are extremely powerful in that they can provide real physical insight into the steady state behavior of a system independently of the details of phenomena occurring within the system. Suppose for example that the steady state is perturbed by small fluctuations δX_α in the thermodynamic forces. According to Eqs. (1.33), the change in entropy production accompanying this perturbation is

$$\Delta\sigma = \tfrac{1}{2} \sum_{\alpha,\beta} \frac{\partial^2\sigma}{\partial X_\alpha \partial X_\beta} \delta X_\alpha \delta X_\beta = \sum_{\alpha,\beta} L_{\alpha\beta} \delta X_\alpha \delta X_\beta = \sum_\alpha \delta J_\alpha \delta X_\alpha \geq 0 \ . \quad (1.34)$$

The inequality, a consequence of properties of the $L_{\alpha\beta}$ which guarantee the second law requirement that Eq. (1.32) be a positive definite quadratic form, implies that small fluctuations around the steady state can only increase σ and therefore cannot lead the system spontaneously to a new steady state. Instead, the system possesses mechanisms with which to damp out the fluctuation and thereby return to the initial state. Alternatively, the minimum entropy production theorem, together with the second law, guarantees that the system will evolve from an arbitrary initial state toward a unique steady state which will then remain stable. In other words, an arbitrary nonequilibrium state will develop in time according to the global relation

$$\frac{dP[S]}{dt} \leq 0, \quad (1.35)$$

the equality representing the steady state situation. Hence, in the linear regime P takes the role of a nonequilibrium thermodynamic potential, providing at the same time a stability criterion for steady states and an evolution criterion for arbitrary nonequilibrium states.

As a consequence of the minimum entropy production principle, it is clear that the appearance of qualitatively new (e.g., highly organized) states is ruled out as long as a physical system remains in the near-equilibrium domain of linear irreversible processes. Nevertheless, under appropriate conditions a steady state having minimum dissipation (i.e., σ_{min}) and therefore maximum efficiency can at the same time possess a low entropy value relative to thermodynamic equilibrium. According to this association of increased complexity (low S) with increased efficiency (σ_{min}) the theorem of minimum entropy production is seen to provide the first link between the concept of "order" and of "evolution" toward a dissipative state. This connection will be made more explicit in our study of nonlinear phenomena.

CHAPTER II: NONLINEAR THERMODYNAMICS

A. Introduction.

Suppose now that the constraints on a physical system are gradually increased so that linear phenomenological laws are no longer strictly valid. (An immediate consequence of nonlinearity in a system is that more than one realizable steady state may be compatible with a given set of constraints.) It is then necessary to consider individually the various types of processes which may occur: (a) Transport processes. In this case linear laws can constitute a good approximation even for quite large gradients (e.g., Fick's diffusion law). (b) Chemical reactions. Rates are related to affinities through $w \sim (1-e^{-A/RT})$, so that linear laws are good only for $A \ll RT$ (R is the gas constant). It is noteworthy that this inequality is frequently not satisfied even very near equilibrium, so that linear laws are seldom adequate for chemical reactions under nonequilibrium conditions. (c) Hydrodynamic effects. As was noted earlier, hydrodynamic systems are always nonlinear due to the explicit nonlinearity of the momentum balance equation.* In summary, then, a large number of important phenomena occurring frequently in physical systems are incapable of even an approximate description in terms of linear irreversible thermodynamics.

According to the last paragraph of Chapter I, attempts to extend the local thermodynamic theory to nonlinear systems can take any of several directions. The earliest success involved formulation by Glansdorff and Prigogine of an evolution criterion appropriate to situations arbitrarily far from equilibrium [7]

$$\frac{d\Phi}{dt} \leq 0 \ . \tag{2.1}$$

Unlike its near-equilibrium counterpart [Eq. (1.35)], however, the evolution criterion could not be related in general to the derivative of a state function (see discussion in Sec. II-B-2). This nonexistence of a general thermodynamic potential implied that the stability of steady states satisfying the equality in Eq. (2.1) could no longer be guaranteed. Hence it became necessary both to search independently for a variational principle which would determine the steady states and at the same time to obtain a separate stability criterion for those states. Before proceeding in the following sections to some details of the formulation by Glansdorff and Prigogine [2], we remark briefly on the context of our interest in these developments.

Aside from hydrodynamic systems, for which unstable transitions to ordered states have long been known, the most likely candidate to exhibit "interesting" far-from-equilibrium behavior appears to be (nonlinear) systems of chemical reactions. Moreover, systems of coupled (bio)chemical reactions under open system conditions are common dissipative elements of life processes in biological systems. For these reasons we specialize now to open systems at mechanical equilibrium.

---

*Hence mechanical equilibrium is an essential feature of steady states corresponding to linear near-equilibrium processes.

Generalization of the principal results to hydrodynamic systems is immediate, and is discussed in detail in Ref. [2].

B. Open Systems at Mechanical Equilibrium.

1. General Comments.

We consider an open, multicomponent mixture of chemical species j among which a number of reactions r may occur. The system is at mechanical equilibrium and for simplicity is maintained isothermal in the absence of external forces. If ρ_j denotes the concentration of species j, then the mass balance equation (1.7) takes the form

$$\frac{\partial \rho_j}{\partial t} = \sum_r \nu_{jr} w_r - \text{div}\,(\rho_j \underset{\sim}{\Delta}_j) \; , \qquad (2.2)$$

where convective effects have been omitted. For our purposes the system is an ideal (dilute) mixture for which a linear diffusion law (of the Fick type) is an excellent approximation:

$$\rho_j \underset{\sim}{\Delta}_j = -\sum_k D_{kj} \,\text{grad}\, \rho_k \; . \qquad (2.3)$$

(Although for the simple applications which we consider diffusive coupling will be ignored, here for completeness the diffusion coefficient matrix is not required to be diagonal.) For chemical reactions, however, linear laws will not at all be applicable, particularly far from equilibrium, so that the various mass balance equations may be nonlinear separately and will in general be nonlinearly coupled through the rates. It is these nonlinearities which will be responsible for qualitatively new phenomena arising far from thermal equilibrium.

2. Evolution Criterion.

Once appropriate chemical kinetic relations are provided, the evolution of the system is specified completely by Eqs. (2.2) together with the phenomenological laws (2.3). From a thermodynamic point of view we consider instead the entropy production [from Eq. (1.28)]

$$P = \int dV \{\sum_r w_r \frac{A_r}{T} - \sum_j \rho_j \underset{\sim}{\Delta}_j \cdot \text{grad}(\frac{\mu_j}{T})\} \geq 0 \; . \qquad (2.4)$$

Since μ_j is now the <u>molar</u> chemical potential, the chemical affinity is defined by Eq. (1.29) with the M_j deleted.

Far from equilibrium the theorem of minimum entropy production is no longer valid. Rather than studying the quantity dP/dt, which is not now expected to provide information concerning the approach to a steady state, we consider separately the contributions due to the changing forces and flows:

$$\frac{dP}{dt} = \frac{d_J P}{dt} + \frac{d_X P}{dt} \equiv \sum_\alpha \{\frac{dJ_\alpha}{dt} X_\alpha + J_\alpha \frac{dX_\alpha}{dt}\} . \qquad (2.5)$$

Specifically, using Eq. (2.4), we evaluate

$$\frac{d_X P}{dt} = \int dV\{\sum_r w_r \frac{d(A_r/T)}{dt} - \sum_j \rho_j \underset{\sim}{\Delta}_j \cdot \frac{d}{dt} \operatorname{grad}(\mu_j/T)\} . \qquad (2.6)$$

After substitution for A_r from Eq. (1.29) and an integration by parts, this expression becomes, for <u>time-independent boundary conditions</u>,

$$\frac{d_X P}{dt} = -\int dV \sum_j [\sum_r \nu_{jr} w_r - \operatorname{div}(\rho_j \underset{\sim}{\Delta}_j)] \frac{d}{dt} (\mu_j/T) \qquad (2.7)$$

or

$$\frac{d_X P}{dt} = -\int dV \sum_j \frac{d\rho_j}{dt} \frac{d}{dt}(\mu_j/T) , \qquad (2.8)$$

where the last step involves substitution from Eq. (2.2). Finally, since μ_j depends only on the composition variables, it follows that

$$\frac{d}{dt}(\mu_j/T) = \frac{1}{T} \sum_k \frac{\partial \mu_j}{\partial \rho_k} \frac{d\rho_k}{dt} , \qquad (2.9)$$

whence

$$\frac{d_X P}{dt} = -\frac{1}{T} \int dV \sum_{j,k} \frac{\partial \mu_j}{\partial \rho_k} \frac{d\rho_j}{dt} \frac{d\rho_k}{dt} . \qquad (2.10)$$

According to equilibrium stability theory [2], expressions such as the integrand here are non-negative at equilibrium. The assumption of local equilibrium ensures, therefore, that the same inequality holds locally in a globally nonequilibrium system, provided that the equilibrium state is stable (with respect to diffusion, in this case, which means that the possibility of a phase separation is excluded). Substituting the appropriate inequality into Eq. (2.10) provides a global evolution criterion for nonequilibrium systems [2]:

$$\frac{d_X P}{dt} \leq 0 \ . \tag{2.11}$$

It is easy to verify, in the limit of linear irreversible processes and provided that Onsager's reciprocal relations are satisfied, that the universal evolution criterion (2.11) reduces to the theorem of minimum entropy production. In that limit, we get $d_X P = d_J P = \frac{1}{2} dP \leq 0$.

Just as in the linear case, the equality in Eq. (2.11) holds at a steady nonequilibrium state. On the other hand, since it represents only part of the time derivative of a nonequilibrium state function, there is no reason to expect in the general nonlinear case that $d_X P$ by itself retains the properties of a total differential. For this reason the asymptotic stability of a steady state which satisfies the equality in Eq. (2.11) is not ensured *a priori*, and must therefore be determined in a separate investigation involving the properties of that particular steady state. This decoupling of evolution and stability is a new feature appearing in the nonlinear regime, and is of special importance in the present context because it permits the occurence of new types of behavior arising beyond a point of instability of an initial steady nonequilibrium state.

3. Thermodynamic Stability Conditions Far From Equilibrium.

In order to illustrate the problem of stability determination for arbitrary nonequilibrium states, we consider briefly near-equilibrium situations for which the minimum entropy production theorem remains valid. In this case one has already $P>0$, $\frac{dP}{dt} \leq 0$, and in addition may write the excess entropy production for fluctuations around the steady state [see Eq. (1.34)]

$$\Delta P \equiv P - P_{\text{steady state}} = \int dV \sum_{\alpha,\beta} L_{\alpha\beta} \delta X_\alpha \delta X_\beta \geq 0 \ , \tag{2.12}$$

at the same time that inequality (1.35) implies

$$\frac{d \ \Delta P}{dt} \leq 0 \ . \tag{2.13}$$

With these two relations the thermodynamic stability problem is completely solved (see Sec. I-G for details). In the language of stability theory, therefore, ΔP is a Lyapounov function. For a wide class of nonlinear differential equations, including the type of evolution equations which govern most thermodynamic systems, it is possible to prove several fundamental theorems, due to Lyapounov, which provide sufficient conditions for stability. Of direct relevance here is the following: The steady state ($\delta X_\alpha = 0$) is stable in a domain D if one can determine a definite function V (e.g., P or ΔP) whose derivative along the motion is either semidefinite with sign opposite to V or vanishes identically in D. Asymptotic stability is ensured if the indicated derivative of V is definite. Since the equality in Eqs. (2.12) and (2.13) hold only at the steady state, the stronger statement applies. Hence in the linear regime the steady state is asymptotically stable.

Is it possible to extend these results beyond the linear domain of irreversible processes? For situations in which the evolution criteria can be transformed into the derivative of a state function, ($\Phi \geq 0$, $d\Phi/dt \leq 0$) the extension is immediate. In the event that such a transformation is not possible (usually the case for nonlinear chemical reactions), a definite answer may still be obtained for stability of steady states with respect to small perturbations [2].

The excess entropy around a reference (steady) state may be written (subscript zero implies evaluation at the reference state)

$$\Delta S \equiv S - S_0 = \delta S + \tfrac{1}{2}\delta^2 S + \ldots , \qquad (2.14)$$

where higher order terms in the expansion are neglected for small deviations ΔS. In general $\delta S \neq 0$ for a state far from equilibrium, and has no well-defined sign. On the other hand $\delta^2 S$ has the remarkable property

$$T\delta^2 S = -\int dV \sum_{j,k} \left(\frac{\partial \mu_j}{\partial \rho_k}\right)_0 \delta\rho_j \delta\rho_k \leq 0 . \qquad (2.15)$$

Here (for convenience) further specialization to a homogeneous reaction mixture has been made (although the inequality may be proven in general [2]) and the inequality established by the same arguments which preceded Eq. (2.11). If $\delta^2 S$ is taken as a Lyapounov function, then the reference state will be asymptotically stable _with respect to infinitesimal perturbation_ whenever

$$\frac{\partial}{\partial t}(\delta^2 S) > 0 \ . \tag{2.16}$$

Eqs. (2.15) and (2.16) provide _sufficient_ conditions for infinitesimal stability.

By differentiating Eq. (2.15) and taking to account the balance equation (2.2) (in the absence of diffusion), an explicit stability criterion for a uniform reacting mixture is obtained:

$$\begin{aligned} T\frac{\partial}{\partial t}(\tfrac{1}{2}\delta^2 S) &= -\int dV \sum_{j,k} \left(\frac{\partial \mu_j}{\partial \rho_k}\right)_o \delta\rho_j \frac{\partial \delta\rho_k}{\partial t} \\ &= -\int dV \sum_{j,k,r} \left(\frac{\partial \mu_j}{\partial \rho_k}\right)_o \nu_{kr} \delta\rho_j \delta w_r \\ &= -\int dV \sum_{k,r} \delta\mu_k \nu_{kr} \delta w_r \\ &= -\int dV \sum_r \delta w_r \delta A_r \geq 0, \text{ for stability.} \end{aligned} \tag{2.17}$$

Finally, it is instructive to point out the close connection between the stability and evolution criteria which have been formulated for arbitrary nonequilibrium states. Indeed, it is possible to obtain the infinitesimal stability condition (2.17) directly from the corresponding variation of the entropy production in Eqs. (2.11) [see Eq. (2.6)]

$$T d_x P = \int dV \sum_r w_r dA_r \leq 0 \ . \tag{2.18}$$

If instead of the _evolution_ $d_x P$ we consider the variation $\delta_x P$ due to a (small) _fluctuation_ around the steady state, the appropriate inequality is

$$T\delta_x P = \int dV \sum_r w_r \delta A_r \geq 0 \ . \tag{2.19}$$

It is always possible to choose the independent affinities (for fixed boundary conditions) such that the corresponding coefficients appearing in Eq. (2.19) vanish at the steady state. (These new coefficients are simply the overall rates of change $\dot{\rho}_j$ in the individual concentration variables.) Expanding these new rates around the steady state gives then, for small fluctuations,

$$T\delta_x P = \int dV \sum_r \delta w_r \delta A_r \geq 0, \tag{2.20}$$

which is precisely the stability criterion (2.17).

It is clear that stability for the steady reference state will be compromised whenever negative terms appear in Eq. (2.17) which can change the sign of the sum. (It is easy to verify that this can never happen in the linear regime.) An important feature of the thermodynamic formulation of stability theory is that it affords an explicit identification of destabilizing contributions to the sum in Eq. (2.17) in terms of thermodynamic quantities. As a consequence one may interpret effects involving unstable transitions (in the following chapters) in a much more intuitive and fundamental manner than is possible in a purely kinetic stability analysis (next section).

C. Stability Theory.

1. Mathematical Stability Theory of Nonlinear Differential Equations.

In order to make explicit the connection between the thermodynamic stability theory developed in preceding sections and more familiar analytical techniques, we now review briefly relevant results from the theory of differential equations. Referring to the full reaction-diffusion system of Sec. II-B-1, we consider the behavior of the system in the neighborhood of a given reference state. That state may be a steady uniform state, a steady inhomogeneous state, or even a time-dependent state. If the symbols $X_j^o(\underline{r},t)$ denote constituent concentrations at the (steady) reference state, then perturbed states in its immediate vicinity may be represented by

$$X_j(\underset{\sim}{r},t) = X_j^o(\underset{\sim}{r},t) + x_j(\underset{\sim}{r},t) \ . \qquad (2.21)$$

Of particular importance from the point of view of stability is the limit of small perturbations x_j ($|x|<<|X^o|$). In this limit the balance equations (2.2) yield a set of linear variational equations

$$\frac{\partial x_j}{\partial t} = \sum_k \left(\frac{\partial F_j}{\partial X_k}\right)_o x_k + \text{div}(D_j \text{ grad } x_j) \ , \qquad (2.22)$$

where $F_j \equiv \sum_r \nu_{jr} w_r$ defines the reaction contribution to Eq. (2.2), and for simplicity the diffusion constant matrix in Eq. (2.3) is assumed diagonal. The great significance of this system of linearized equations lies in the equivalence (in many cases) of its stability properties to those of the corresponding nonlinear system. Application of the methods of Lyapounov (see preceding subsection) has been discussed in the current context by Nicolis [8]. Here we summarize a few of the principal results of an analysis in terms of normal modes, and refer to Ref. [8] for details and additional literature citations.

Consider now infinitesimal perturbations of the form (in 1 dimension)

$$x_j(\underset{\sim}{r},t) = x_j^o \exp(\omega t + ir/\lambda) \ , \qquad (2.23)$$

having frequency ω and wavelength λ. If periodic boundary conditions are applied, and D is assumed constant, then substitution of Eq. (2.23) into Eq. (2.22) gives the characteristic equation

$$\left|(\omega + \frac{D}{\lambda^2})\delta_{jk} - \left(\frac{\partial F_j}{\partial X_k}\right)_o\right| = 0 \qquad (2.24)$$

which must be satisfied for the system (2.22) to possess a non-trivial solution. The eigenfrequencies ω are in general complex, $\omega \equiv \omega_r + i\omega_i$, the stability properties of the system of equations (2.22) depending on the sign of the real part, ω_r. We distinguish among several possibilities: (1) All $\omega_r<0$: the system (2.22) is stable; all normal modes decay. (2) At least one $\omega_r>0$: the system is unstable; the mode with $\omega_r>0$ will grow. In either case the behavior of individual modes will be oscillatory if the corresponding $\omega_i \neq 0$, and monotonic otherwise. (3) At least one $\omega_r=0$, all other $\omega_r<0$: the system is said to be marginally stable with respect to the mode having $\omega_r=0$; that mode will neither decay nor grow, but may oscillate around the steady reference state if it has in addition $\omega_i \neq 0$.

To this point nothing has been said about the role of diffusion in the stability analysis. Generally one expects the presence of diffusion to have a stabilizing effect (see discussion in next chapter). Diffusion has the additional property, however, by permitting consideration of spatially inhomogeneous perturbations as well, of increasing enormously the variety of (possibly unstable) situations which may arise. This important subject will be explored further in connection with specific models in the following chapter.

Before turning to a thermodynamic discussion of the behavior of normal modes near steady states, we remark briefly on the connection between infinitesimal (linear) stability and global stability of the complete nonlinear system (2.2). By means of methods due to Lyapounov (Sec. II-B-3 and Ref. [8]), it is possible to establish the following general results [9]:

(1) If all roots of the characteristic equation (2.24) have negative real parts, the steady state is asymptotically stable for Eq. (2.2), whatever the nonlinear terms.
(2) If at least one characteristic value has a positive real part, the steady state is unstable whatever the nonlinear terms.
(3) If the characteristic equation has roots with zero real parts, then the steady state may be stable or unstable depending on the explicit form of the nonlinear terms.

In terms of these stability conditions, then, the specific chemical nonlinearities in Eqs. (2.2) are seen to influence the stability properties of the steady state solutions only in the case of marginal stability.

2. Thermodynamic Interpretation.

As we have seen, a positive characteristic value for a single normal mode is sufficient to compromise the stability of (stationary) solutions to the system (2.2). Consequently we look now from a thermodynamic viewpoint at the behavior of a single normal mode around a steady nonequilibrium state. As in the previous subsection, infinitesimal perturbations (in thermodynamic variables) take the form

$$\delta w_r = \delta w_r^o \exp[(\omega_1+i\omega_2)t] \quad , \quad \delta A_r = \delta A_r^o \exp[(\omega_1+i\omega_2)t] \ , \quad (2.25)$$

where for convenience spatial uniformity is assumed.

Because of the appearance of complex quantities in the normal mode representation, it is necessary to reformulate slightly the evolution and stability criteria so that thermodynamic quantities remain real. Hence Eq. (2.6) is rewritten, in expanded form about a steady state,

$$\frac{Td_xP}{dt} = \frac{Td}{dt}\delta_xP = \tfrac{1}{2}\int dV\sum_r\{\delta w_r^* \frac{d}{dt}\delta A_r + \delta w_r \frac{d}{dt}\delta A_r^*\} \ , \quad (2.26)$$

where the superscript * implies complex conjugation. Explicitly introducing the perturbations (2.25) gives

$$\frac{Td}{dt}\delta_xP = \tfrac{1}{2}\int dV\sum_r\{\omega_1(\delta w_r^*\delta A_r + \delta w_r\delta A_r^*) + i\omega_2(\delta w_r^*\delta A_r - \delta w_r\delta A_r^*)\} \leq 0. \quad (2.27)$$

The coefficient of ω_1 is just the <u>excess entropy production</u>, δP, associated with the normal mode. If the excess entropy (2.15) is written in the complex form,

$$T\delta^2S = -\tfrac{1}{2}\int dV \sum_{j,k} \left(\frac{\partial\mu_j}{\partial\rho_k}\right)_o(\delta\rho_j^*\delta\rho_k + \delta\rho_j\delta\rho_k^*) \ , \quad (2.28)$$

then the result analogous to the stability criterion (2.17) is

$$\delta P = \frac{1}{2T}\int dV\sum_r(\delta w_r^*\delta A_r + \delta w_r\delta A_r^*) = \frac{\partial}{\partial t}(\delta^2S) \geq 0 \ . \quad (2.29)$$

Similarly, the coefficient of ω_2 is a real quantity of thermodynamic significance:

$$\delta\Pi \equiv \frac{i}{2T}\int dV\sum_r(\delta w_r^*\delta A_r - \delta w_r\delta A_r^*) \ . \quad (2.30)$$

For a single normal mode it is easy to verify that

$$\omega_1 \delta P = \omega_1 \frac{\partial}{\partial t}(\delta^2 S) = \omega_1^2 \delta^2 S \leq 0 \ , \qquad (2.31)$$

and

$$\omega_2 \delta \Pi = \frac{Td}{dt}\delta_x P - \omega_1 \delta P$$

$$= (\omega_1^2 + \omega_2^2)\delta^2 S - \omega_1^2 \delta^2 S = \omega_2^2 \delta^2 S \leq 0 \ . \qquad (2.32)$$

Inequality (2.31) provides a stability criterion for a single normal mode, combining in a single expression the heretofore distinct criteria $\delta P>0$ [Eq. (2.16)] and $\omega_1<0$. At the same time the equality defines thermodynamically the meaning of a state of marginal stability, the transition point $\delta P=0=\omega_1$ separating regions of stability and instability for the normal mode.

The second inequality (2.32) relates the oscillatory frequency ω_2 to the quantity $\delta\Pi$, assigning in thermodynamic terms the direction of rotation around the steady state. It too provides a thermodynamic transition, at points for which $\delta\Pi$ (and ω_2) become nonzero, between aperiodic and periodic motion near the steady state.

If the preceding equations for δP and $\delta\Pi$ are made explicit by expanding the excess reaction rates δw_r in terms of the excess forces $\frac{1}{T}\delta A_r$, then one obtains the important result that $\delta\Pi$ (and therefore also ω_2) always vanishes near equilibrium, while $\delta P>0$ (and ω_1) can never vanish in that region as long as the theorem of minimum entropy production remains valid. In accord with the conclusions of Sec. I-G, therefore, the appearance of oscillations and instabilities is exclusively a feature of nonlinear systems under far-from-equilibrium conditions.

CHAPTER III: DISSIPATIVE STRUCTURES

A. The Thermodynamic Problem. Equilibrium vs Nonequilibrium Order.

For a system in thermodynamic equilibrium, instabilities can occur only at phase transition points. In this case beyond instability there appears a qualitatively new phase, which may correspond to a more ordered state (e.g., liquid⟶solid or paramagnet⟶ferromagnet). The important point here is that once such a new structure is formed, its maintenance does not require any interaction with the outside world. In other words, (in a constant environment) equilibrium structures are self-sustaining.

As was seen in the last chapter, away from equilibrium a new type of instability may occur as a direct consequence of the same constraints which are responsible for maintaining a steady nonequilibrium state. Unlike their equilibrium counterpart, however, such nonequilibrium instabilities are by no means universal. Indeed, it is not difficult to demonstrate that the stability criterion (2.16) for nonequilibrium states may be violated only for a certain class of nonlinear systems which are in addition maintained beyond a critical distance from thermodynamic equilibrium. Restricting our attention to such systems, then, the question which now concerns us is whether beyond instability there may appear ordered structures of a new type. As was noted in the introduction to this review, such structures would differ markedly from the equilibrium variety, depending both for creation and maintenance on a continuous exchange of energy and matter with the surroundings.

In this chapter we investigate a few simple model systems which will illustrate the variety of behavior which may arise beyond instability. Before examining the various possibilities, however, we first formulate the thermodynamic problem in a quite general form. Consider a nonisolated (closed or open) system subject to external constraints which maintain a steady nonequilibrium state. At the steady state the various thermodynamic rates or flows depend on a set $\{X\}$ of independent generalized thermodynamic forces (e.g., temperature gradient, chemical affinity, etc.) which provide in addition a measure of the deviation from equilibrium. We may distinguish three distinct regions characterized by the magnitude of the parameters $\{X\}$: (a) $\{X=0\}$ The system is in thermodynamic equilibrium. The uniqueness and stability of this state are ensured by the existence of a thermodynamic potential--one of the free energies F, G. (b) {$X \neq 0$ but "small"} The equilibrium regime is extended continuously into the near-equilibrium domain. If the equilibrium state is stable, then in the linear region states on this "thermodynamic branch" of solutions are guaranteed unique and stable by the theorem of minimum entropy production. (c) {X "large"} Far from equilibrium, a thermodynamic potential does not exist in general. Therefore the uniqueness of the steady state is no longer guaranteed and, if the system obeys nonlinear kinetic laws, it may exhibit more than one steady state compatible with a given set of boundary conditions. One of these states belongs to the thermodynamic branch, which is continued into the far-from-equilibrium region, but its stability is no longer ensured automatically. Once the thermodynamic branch becomes unstable, the system may undergo a discontinuous transition to a new regime which then becomes stable, giving rise to any of a variety of new effects.

The characteristic pattern of behavior described here has long been a "classical" area of study in hydrodynamics, which abounds with examples of instabilities and subsequent evolution to new ordered regimes. Indeed, it was in applications to well-known hydrodynamic instabilities that many aspects of nonlinear thermodynamics were originally developed by Glansdorff and Prigogine [7]. Only recently has a systematic study of instabilities in purely dissipative systems been undertaken as well.

According to the development in preceding sections, the key point in the thermodynamic analysis of nonequilibrium instabilities is the appearance of negative terms in the stability criterion (2.16). Under appropriate conditions these negative contributions may dominate and the stability of the reference state is then compromised. As we shall see more clearly in this chapter, there are many ways in which chemical reactions may produce negative contributions to the stability criterion.

In contrast, the Navier-Stokes equations of fluid dynamics assume a universal form, with a limited number of dimensionless parameters which determine the stability properties of hydrodynamic systems. Finally, we recall that all such instabilities can arise only at a finite distance from thermodynamic equilibrium: that is, their occurrence requires a minimum level of dissipation. For this reason Prigogine has coined the term "dissipative structures" to distinguish the new types of ordered regimes arising beyond nonequilibrium transitions [2].

B. Symmetry-Breaking Instabilities.*

1. A Simple Model.

In this section we outline the general features of a simple chemical model, specializing in following sections to particular operating regimes in which various types of ordered behavior are found. Consider the following reaction scheme:

$$\begin{aligned} A &\underset{k_{-1}}{\overset{k_1}{\rightleftharpoons}} X \\ B + X &\underset{k_{-2}}{\overset{k_2}{\rightleftharpoons}} Y + D \\ 2X + Y &\underset{k_{-3}}{\overset{k_3}{\rightleftharpoons}} 3X \\ X &\underset{k_{-4}}{\overset{k_4}{\rightleftharpoons}} E\ . \end{aligned} \tag{3.1}$$

The autocatalytic third step is a convenient way to introduce nonlinearity. Since the trimolecular reaction is an unlikely step, however, this mechanism is taken as a convenient model rather than as a representation of an actual chemical process. The net reaction is $A + B \rightleftharpoons D + E$, and is the sum of two distinct transformations, $A \rightleftharpoons E$ and $B \rightleftharpoons D$. If an ideal (dilute) reaction mixture is as-

*The terminology as used here refers to temporal as well as spatial symmetry.

sumed, then the overall chemical affinity (in units of RT) for the scheme (3.1) is

$$A = \ln \frac{k_1 k_2 k_3 k_4}{k_{-1} k_{-2} k_{-3} k_{-4}} \frac{AB}{ED} , \tag{3.2}$$

where the symbol A denotes the concentration of species A, etc.

The system is open to reservoirs of initial reactants A, B and final products D, E, the concentrations of which are maintained constant in space and time. For simplicity we consider diffusion along a single space coordinate r. Then the chemical kinetic equations representing the system (3.1) are [c.f., Eqs. (2.2) and (2.3)]

$$\frac{\partial X}{\partial t} = k_1 A - k_2 X - k_2 BX + k_{-2} DY + k_3 X^2 Y - k_{-3} X^3 + k_4 X - k_{-4} E + D_X \frac{\partial^2 X}{\partial r^2}$$

$$\frac{\partial Y}{\partial t} = k_2 BX - k_{-2} DY - k_3 X^2 Y + k_{-3} X^3 + D_Y \frac{\partial^2 Y}{\partial r^2} \quad . \tag{3.3}$$

Here D_x, D_y are the (constant) diffusion coefficients for intermediates X, Y.

Our initial interest in these equations is the stability properties of the thermodynamic branch of stationary solutions in various regions characterized by the magnitude of the overall affinity (3.2). If the time derivatives are set to zero in Eqs. (3.3), then it is not difficult to see that this system admits a single spatially homogeneous steady state solution

$$X_o = \frac{k_1 A + k_{-4} E}{k_{-1} + k_4} \quad , \quad Y_o = X_o \frac{k_2 B + k_{-3} X_o^2}{k_{-2} D + k_3 X_o^2} \quad , \tag{3.4}$$

which for all values of the reservoir concentrations constitutes the thermodynamic branch for the scheme (3.1). In particular the state of thermodynamic equilibrium corresponds to

$$\frac{k_1 k_4}{k_{-1} k_{-4}} \frac{A_{eq}}{E_{eq}} = \frac{k_2 k_3}{k_{-2} k_{-3}} \frac{B_{eq}}{D_{eq}} = 1 , \tag{3.5}$$

which implies

$$X_{eq} = \frac{k_1}{k_{-1}} A_{eq} \quad , \quad Y_{eq} = \frac{k_{-3}}{k_3} X_{eq} = \frac{k_1 k_{-3}}{k_{-1} k_3} A_{eq} \quad . \tag{3.6}$$

One knows from general thermodynamic arguments that the homogeneous steady state (3.4) is stable at or near equilibrium. Therefore we turn now to a study of this thermodynamic branch in the far-from-equilibrium domain.

In order to simplify the anyalysis, but without loss of any essential features, we take the limit D→0, E→0, which means physically that these products are removed from the "vicinity" of the reaction as soon as they are produced. The corresponding reactions become irreversible, and the affinity (3.2) tends to infinity in this limit. This means that the system operates at an "infinite distance" from equilibrium. Hence there is no qualitative change if the reverse reactions in the remaining two steps are neglected as well. The system is now completely irreversible, but it is still possible to define the continuation of the thermodynamic branch (3.4).

The simpler system of kinetic equations which results is obtained by setting all rate constants for the reverse reactions to zero. In this case Eqs. (3.3) become

$$\begin{aligned} \frac{\partial X}{\partial t} &= k_1 A - k_2 BX + k_3 X^2 Y - k_4 X + D_x \frac{\partial^2 X}{\partial r^2} \\ \frac{\partial Y}{\partial t} &= k_2 BX - k_3 X^2 Y + D_x \frac{\partial^2 Y}{\partial r^2} \quad , \end{aligned} \tag{3.7}$$

with homogeneous steady state solution, from Eqs. (3.4),

$$X_o = \frac{k_1}{k_4} A_o \quad , \qquad Y_o = \frac{k_2 B}{k_3 X_o} = \frac{k_2 k_4 B}{k_1 k_3 A} \quad . \tag{3.8}$$

The stability of this solution is examined first via a normal mode analysis. If (small) perturbations of the form (2.23) are considered, then the linear variational equations (2.22) become

$$\begin{aligned} \omega x &= (-k_2 B + 2k_3 X_o Y_o - k_4)x + k_3 X_o^2 y - \frac{D_x}{\lambda^2} x \\ \omega y &= (k_2 B - 2k_3 X_o Y_o)x - k_3 X_o^2 y - \frac{D_x}{\lambda^2} y \ , \end{aligned} \tag{3.9}$$

and a little algebra yields from Eq. (2.24) a dispersion relation giving $\omega(\lambda)$:

$$\omega^2 + (-k_2B + k_3X_o^2 + k_4 + \frac{D_x+D_y}{\lambda^2})\omega - k_2B\frac{D_y}{\lambda^2} + (k_3X_o^2 + \frac{D_y}{\lambda^2})(k_4 + \frac{D_x}{\lambda^2}) = 0. \quad (3.10)$$

It is a simple matter now to find the points of marginal stability (i.e., $\omega_{real}=0$) separating regions of stability and instability with respect to the various normal modes. Suppose B is taken as the single independent variable (for fixed A, D_x, D_y). As B increases, there are two points at which ω_r may vanish:

$$B' = \frac{1}{k_2}(k_3X_o^2 + k_4 + \frac{D_x+D_y}{\lambda^2}) \text{ and } B'' = \frac{\lambda^2}{k_2D_y}(k_3X_o^2 + \frac{D_y}{\lambda^2})(k_4 + \frac{D_x}{\lambda^2}). \quad (3.11)$$

At B=B', the coefficient of ω^1 in Eq. (3.9) vanishes, and both roots are either purely imaginary, if the constant term (coefficient of ω^0) is positive, or real and nonzero if the constant term is negative. At B=B'', the constant term is zero, so that one root vanishes identically, while the other is real and non-zero. There are then two types of situations which may arise, depending, for fixed A, D_x, D_y (and rate constants), on the wavelength of the disturbance:

(a) B'<B'': For B<B', the homogeneous steady state (3.8) is stable, the decay of perturbations becoming oscillatory as B-→B'. The point B' is an oscillatory marginal state. For B>B', therefore, an initial perturbation will rotate around the steady state as it grows. At some point in B'<B<B'', both complex roots (with $\omega_r>0$) become real. Beyond that point all initial perturbations would grow monotonically. Passage through the point B=B'' would have no effect on the stability of the solution (3.8), as one root would remain positive and real.

(b) B'>B'': In this case, the first point of instability, B'', corresponds to a nonoscillatory marginal state. For B>B'', both roots are real, and one is always positive. Therefore all initial perturbations grow monotonically in this region. (The marginal state at B' does not exist, since the constant term in Eq. (3.8) is negative once B>B''.)

In summary, then, as B is increased continuously, the homogeneous thermodynamic branch (3.8) will experience an oscillatory instability at B' if B'<B'' or a nonoscillatory instability at B'' if B'>B''.

In order to understand better the nature of the instabilities possible at these two points, consider the "critical" wavelengths for disturbances which will produce instability at the smallest value of B. Minimizing the expressions in Eqs. (3.11) with respect to λ yields

$$\lambda_c' \longrightarrow \infty \qquad B_c' = \frac{k_3}{k_2} X_o^2 + \frac{k_4}{k_2} \tag{3.12a}$$

and

$$\lambda_c'' = \left(\frac{D_x D_y}{k_3 k_4 X_o^2}\right)^{1/4} , \qquad B_c'' = \left[\left(\frac{k_3 D_x}{k_2 D_y}\right)^{1/2} X_o + \left(\frac{k_4}{k_2}\right)^{1/2}\right]^2 . \tag{3.12b}$$

In an actual chemical mixture there will usually be present spontaneous fluctuations having a wide range of wavelengths. As B increases, then, the instability which appears will correspond to the smaller of the two critical values B_c. For case (a) above Eqs. (3.12a) imply that very long wavelength perturbations will grow and oscillate, the system remaining homogeneous. For case (b), however, the possibility of appearance of finite inhomogeneities is seen from Eqs. (3.12b) to involve a cooperative effect between chemical reactions and diffusion. Indeed, whenever the rates of diffusion and chemical reactions are "comparable", short wavelength inhomogeneities may grow. As diffusion rates become dominant, however, the instability occurs for increasing wavelengths, and the system may again remain nearly homogeneous.

Which type of instability will occur is seen from Eqs. (3.12) to depend on the ratio of the diffusion coefficients. In fact, the possible emergence of a spatial inhomogeneity requires $B_c'' < B_c'$, or (for simplicity taking all $k_i = 1$)

$$\left(\frac{D_x}{D_y}\right)^{1/2} < \frac{1}{A}\left[-1 + (B_c')^{1/2}\right] . \tag{3.13}$$

This inequality will hold only for sufficiently small D_x/D_y, and in particular cannot be satisfied for <u>equal</u> diffusion coefficients.*

---

*Unequal diffusion coefficients is a common, although not universal, prerequisite for spatial symmetry breaking.

For a deeper insight into the role of diffusion and of the various chemical reactions in the onset of instabilities, we now turn to a thermodynamic stability analysis of scheme (3.1). To compute the excess reaction rates δw and affinities δA corresponding to a small deviation (x,y) from the steady state (X_o, Y_o), first calculate these quantities for the full system (3.1), then take the limit in which reverse reactions are neglected:*

$$\begin{aligned}
\delta w_1 &= 0 & \tfrac{1}{T}\delta A_1 &= -\frac{x}{X_o} \\
\delta w_2 &= k_2 B x & \tfrac{1}{T}\delta A_2 &= \frac{x}{X_o} - \frac{y}{Y_o} \\
\delta w_3 &= 2k_3 X_o Y_o x + k_3 X_o^2 y & \tfrac{1}{T}\delta A_3 &= -\frac{x}{X_o} + \frac{y}{Y_o} \\
\delta w_4 &= k_4 x & \tfrac{1}{T}\delta A_4 &= \frac{x}{X_o}
\end{aligned} \qquad (3.14)$$

For diffusive flows and forces the excess quantities are, for intermediate X,*

$$\delta(X\Delta_x) = \delta(D_x \text{ grad } X) = D_x \text{ grad } x, \; \delta(\text{grad } \tfrac{\mu}{T}x) = \frac{1}{X_o} \text{ grad } x \;, \qquad (3.15)$$

and similarly for intermediate Y.

The excess entropy production may now be calculated from Eq. (2.29), with diffusion contribution added [c.f., Eq. (2.4)]

$$\begin{aligned}
2\delta P = {} & [(k_2 B - 2k_3 X_o Y_o)x - k_3 X_o^2 y](\frac{x^*}{X_o} - \frac{y^*}{Y_o}) + \frac{k_4}{X_o}xx^* \\
& + \frac{D_x}{X_o}|\text{grad } x|^2 + \frac{D_y}{Y_o}|\text{grad } y|^2 + \text{complex conjugate},
\end{aligned} \qquad (3.16)$$

where an integration over volume is implied. If the steady state concentrations are substituted from Eqs. (3.8), then the cross terms in xy^* and x^*y are seen to cancel with each other's complex conjugates, leaving

$$2\delta P = (k_2 B - 2k_3 X_o Y_o + k_4)\frac{xx^*}{X_o} + \frac{k_3 X_o^2}{Y_o} yy^* + \frac{D_x}{X_o}|\text{grad } x|^2 + \frac{D_y}{Y_o}|\text{grad } y|^2 + \text{c.c.} \qquad (3.17)$$

---

*For simplicity of notation chemical potentials are expressed in units of R (the gas constant); e.g., $\delta(\mu_x/T) = \delta(\ln X) = (\ln x)/X_o$.

or

$$\delta P = (k_2 B - 2k_3 X_o Y_o + k_4 + \frac{D_x}{\lambda^2}) \frac{x_r^2 + x_i^2}{X_o} + (k_3 X_o^2 + \frac{D_y}{\lambda^2}) \frac{y_r^2 + y_i^2}{Y_o} , \quad (3.18)$$

where the explicit form (2.23) for the perturbations (x,y) has been introduced. In this form the role of various processes is especially transparent. As was anticipated earlier, a "dangerous" (negative) contribution to δP is introduced by the autocatalytic third step of the scheme (3.1). In addition, the effect of diffusion is always stabilizing. On the other hand, including diffusion explicitly in the model permits consideration of inhomogeneous systems. Hence the class of possible perturbations which may initiate instabilities is widened enormously by the presence of diffusion. Substituting for (X_o, Y_o) in the remaining terms and recalling in addition the linearized equations (3.9), it is straightforward to verify that δP does indeed vanish at points of marginal stability.

The quantity $\delta\Pi$ [Eq. (2.30)] follows from Eq. (3.6) by subtracting, rather than adding, the complex conjugate expression:

$$\delta\Pi = 2k_3 X_o (x_i y_r - x_r y_i) . \quad (3.19)$$

2. Chemical Oscillations.

Suppose that the "most probable" initial fluctuation is spatially homogeneous. In this case $\lambda \to \infty$ and diffusion plays no role in the evolution of the system. According to Eqs. (3.12), $B_c'' \to \infty$ also and an oscillatory instability will occur at $B = B_c'$. Before the instability point is reached, fluctuations either decay monotonically or perform damped oscillations about the steady state. Beyond instability, fluctuations are amplified, the system leaves the steady state, and evolves to an undamped (homogeneous) periodic regime (shown in Fig. 1). The resulting dissipative structure is a unique (in this example) limit cycle* which

*For a comprehensive mathematical treatment see Ref. [9]. For a discussion of experimental and theoretical studies on chemical oscillations see Refs. [8] and [10].

is asymptotically stable, its characteristic period and amplitude depending only on the values assigned to the various parameters of the system. In more general examples, a discrete set of orbits may be accessible for a single set of system parameters. The particular limit cycle which results depends then on the choice of initial conditions. The coherent behavior implied by the "ergodic" evolution of this system is characteristic of a chemical clock. This contrasts markedly with the typically incoherent behavior of oscillations in conservative systems (e.g., simple pendulum, Volterra-Lotka models for competing populations [8]), for which a continuum of amplitudes and periods is available and the corresponding orbits are not asymptotically stable.

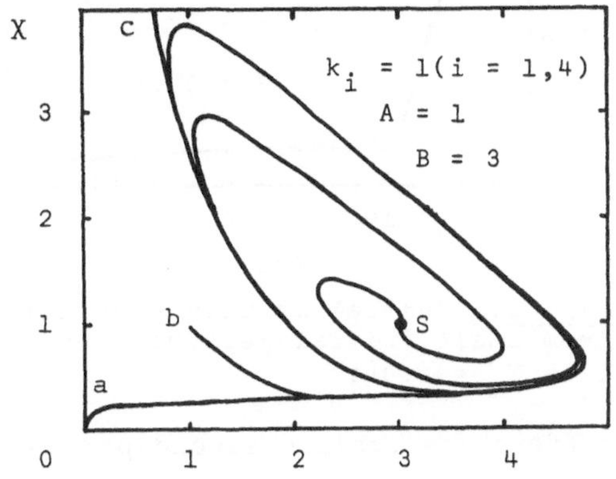

Fig. 1. Approach to stable limit cycle from unstable steady state S and from other initial conditions (a,b,c).

3. Spatial Structure.

Suppose now that the "most probable" initial fluctuation is space-dependent. As B is increased gradually, then, a nonoscillatory instability will occur first at the wavelength λ_c'' for which B" is minimum [provided, of course, that $B'(\lambda_c'') > B_c'' = B''(\lambda_c'')$ in Eqs. (3.11)]. The evolution of the system beyond this instability has been studied numerically. The most striking conclusion of the various computer "experiments" on the scheme (3.1) is that, for $B > B_c''$, there exists a new stable steady state which is spatially inhomogeneous.

In the initial computer simulations a two-box model was adopted for simplicity. Initial and final products are distributed uniformly throughout. Diffusion is represented as a flow of matter across the surface of separation between two boxes each of which is spatially uniform. In this model diffusion is governed by laws in which simple

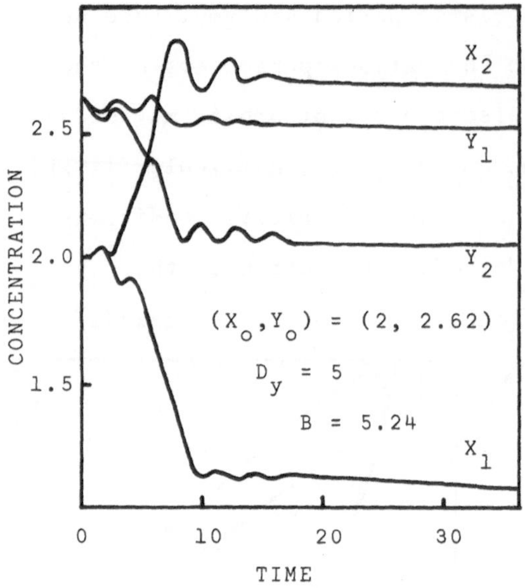

Fig. 2. Spatial structure resulting from small initial perturbation $Y_2 - Y_1 = 0.04$.

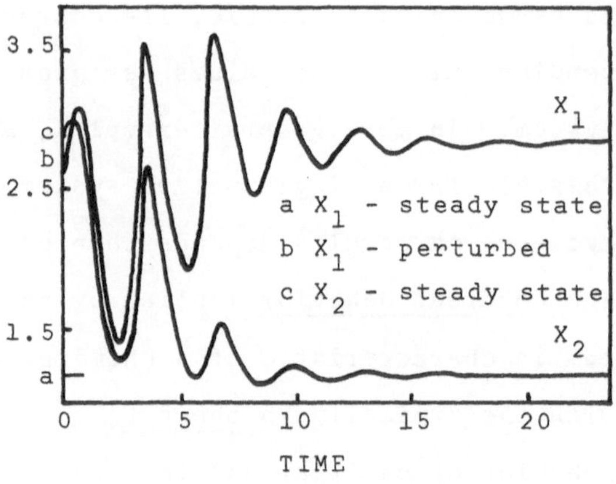

Fig. 3. Exchange of stabilities due to large initial perturbation in $X_1 (X_1 - X_1^o = 2.6 - 1.17 = 1.43)$.

differences replace second partial derivatives (e.g., $\partial^2 X/\partial r^2 \rightarrow X_2 - X_1$). The evolution of each box is then described by a pair of kinetic equations of the form (3.7), the two sets of equations being coupled through the diffusion terms. It is easy to see that this system of four equations admits as a unique homogeneous time-independent solution the steady state (3.8). In addition, however, there exists a spatially inhomogenous solution such that either $(X_1 < X_2 , Y_1 > Y_2)$ or $(X_1 > X_2 , Y_1 < Y_2)$, the model being symmetrical. A typical evolution to such a nonuniform state is displayed in Fig. 2. The spatial structure which results is found numerically to be extremely stable. An initial perturbation which is sufficiently large, however, may lead to an exchange of concentrations in the two-box steady state (Fig. 3). It is important to note that small fluctuations cannot reverse the configuration. Indeed, reversing can occur only for perturbations of the same order of magnitude as the difference in concentration between the boxes. Hence the spatial structure is stable as far as "spontaneous" infinitesimal fluctuations are concerned.

The results of the simple two-box model have been extended to multi-box systems in order to represent diffusion in a more realistic manner and to permit more freedom in the choice of wavelength for the initial perturbation. A typical final state obtained with 100 boxes (for fixed boundary values of X,Y) is shown in Fig. 4. The wavelength of the resulting spatial periodicity is determined approximately by λ_c'' , the wavelength of the fastest-growing fluctuation beyond the instability. In addition, the particular final configuration which arises depends to a degree on the type of initial perturbation. This effect is seen most easily in the two-box model: the final concentration of X will be higher in the box in which the fluctuation induces an increase in the rate of the autocatalytic third step. As a consequence of this <u>primitive memory effect</u> one sees that spatial dissipative structures have a capability for storing "information" accumulated in the past.

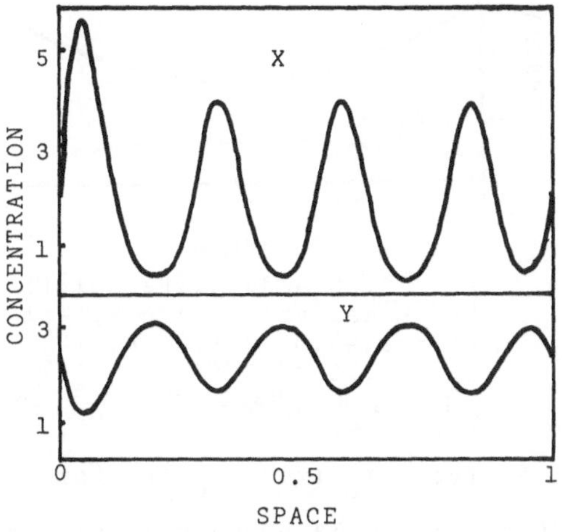

Fig. 4. Stationary distribution for 100 boxes with fixed boundary concentrations (X = 2, Y = 2.62).

Recently Nicolis has verified analytically the existence for this model of time-independent spatially inhomogeneous states [11]. By means of a perturbation calculation which is exact near the marginal state corresponding to B_c'', he has shown also that more than one such state is possible. As was suggested above, in this case the final state which results depends crucially on the type of the initial perturbation.

4. Localized Spatial Structure and Chemical Waves. [12]

If a chemical reaction scheme exhibits finite spatial inhomogeneities for some constituents, then it is natural to expect that the distribution of other participating species may be nonuniform as well. Therefore we consider now a more realistic model for system (3.1) in which the reservoir concentrations also may be space-dependent. If for simplicity we let reactant A be the single additional diffusing species, then the system is described by Eqs. (3.7) together with the kinetic equation for A(r,t),

$$\frac{\partial A}{\partial t} = -k_1 A + D_a \frac{\partial^2 A}{\partial r^2} \quad . \qquad (3.20)$$

This choice yields an especially simple model because A(r,t) satisfies a closed equation (while B, for example, would not). Hence Eq. (3.20) may be solved independently, its solution appearing then as a known function A(r,t) in Eqs. (3.7).

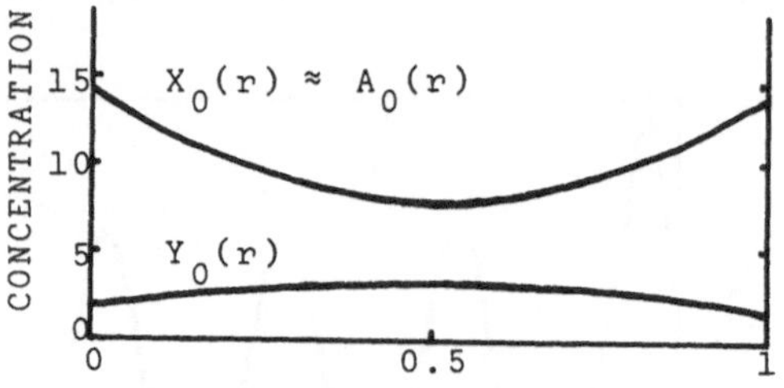

Fig. 5. Steady state solution. Numerical values are $k_i = 1$ (i = 1,4), $D_x = 1.05 \times 10^{-3}$, $D_y = 5.25 \times 10^{-3}$, $D_a = 197 \times 10^{-3}$, B = 26.0, $\bar{X} = \bar{A} = 14.0$, $\bar{Y} = 1.86$.

A typical steady state profile $A_0(r)$, $X_0(r)$, $Y_0(r)$ for fixed values $\bar{A},\bar{X},\bar{Y}$ at the boundaries ($0 \leq r \leq L = 1$) appears in Fig. 5. At each point r the values X_0, Y_0 differ very little from the prescription (3.4) as functions of $A_0(r)$. Hence the effect of diffusion on this new "thermodynamic solution" (reducing to Eqs. (3.8) when $D_A \to \infty$) is seen to be minimal. An infinitesimal stability analysis of the steady state yields a kind of "nonequilibrium phase diagram" which is shown in Fig. 6. In region I, fluctuations decay and the reference state is stable. In the other two regions the steady state is unstable, the growth of fluctuations being monotonic in region II and oscillatory in region III. We now consider separately

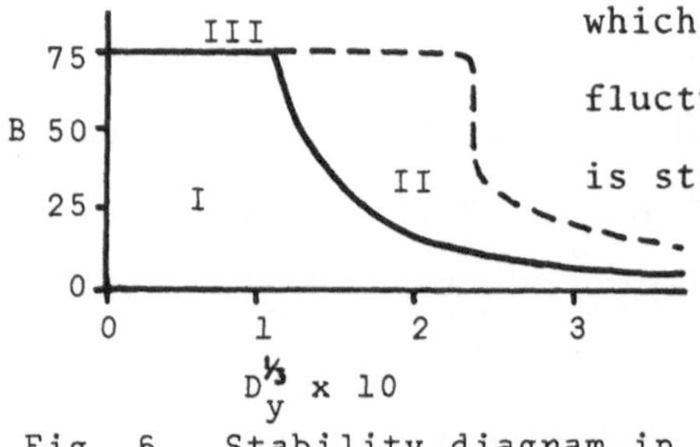

Fig. 6. Stability diagram in the B-D_y plane.

situations in each of the unstable zones.

For the parameter values of Fig. 5, the system is in region II. A numerical integration* of the kinetic equations (for a space grid of 79 points) shows that small perturbations of the steady state are amplified. After some time the system reaches a new steady state (Fig. 7) corresponding to a spatial organization maintained by the flow of matter through the boundaries. In contrast with previous studies (preceding section), here the final state is a localized (spatial) dissipative structure. In fact, the system appears to determine its own "natural" boundaries within which a sharp, short-wavelength structure is established. This remarkable feature is clearly due to the nonuniform distribution of A in the system, and in addition depends crucially on the values assigned to B and D_y. On one hand, with increasing B the structure spreads until, when B exceeds the critical value B" [Eqs. (3.11)] corresponding to the largest value of A in the system ($\bar{A}$, at the boundary), it fills the entire space. On the other hand, we expect from Eq. (3.13) that the spatial structure will shrink and finally disappear as D_y decreases toward D_x. Finally, questions of stability and uniqueness of such structures lead, as was hinted for the two-box model, to some most interesting possibilities. In the present case, and more generally in the limit of a continuous system, the problem of additional structures compatible with given constraints is likely to be more complex. In a real system the evolution to a "final" state might consist of a succession of events,

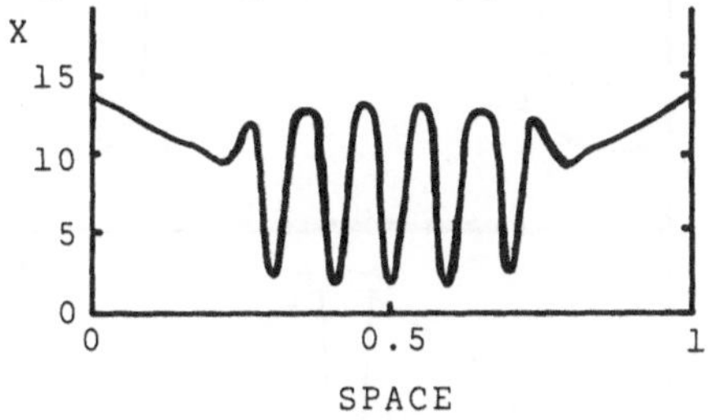

Fig. 7. Localized spatial dissipative structure

* In the integration A(r,t) is replaced by the steady state distribution $A_o(r)$. This approximation is justified since A_o is a stable configuration and since the choice $D_a >> D_x, D_y$ implies a rapid relaxation to the profile A_o.

each beginning with an instability (due, for example, to changing surroundings) followed by a "choice" among various accessible states. The conjecture is then that the history of such systems, as manifested in the "most probable" type of fluctuation to occur in a particular environment, plays a prominent role in subsequent evolution.*

Looking now at region III, we let B=77.0 and D_y=0.66x10^{-3}, all other parameters remaining unchanged. Again the unperturbed steady state has the form given in Fig. 5. Integrating the kinetic equations beyond instability (for a space grid of 40 points) yields a dissipative structure which is inhomogeneous in time as well as space. The "final" regime corresponds to the periodic (in time) sequence shown in Fig. 8. From an initial profile (e.g., Fig. 5) a well is formed in the middle and propagates outward. After a short time propagation stops and a slow buildup in X takes place at two points (9 and 31). When X reaches a maximum at these points, an extremely rapid inward wave begins. After the two wavefronts "collide", X decreases slowly to its initial profile from which a new cycle begins. As in region II, the system is seen to establish its own boundaries between which it exhibits a wave-like solution. Within its "interior" (i.e.,

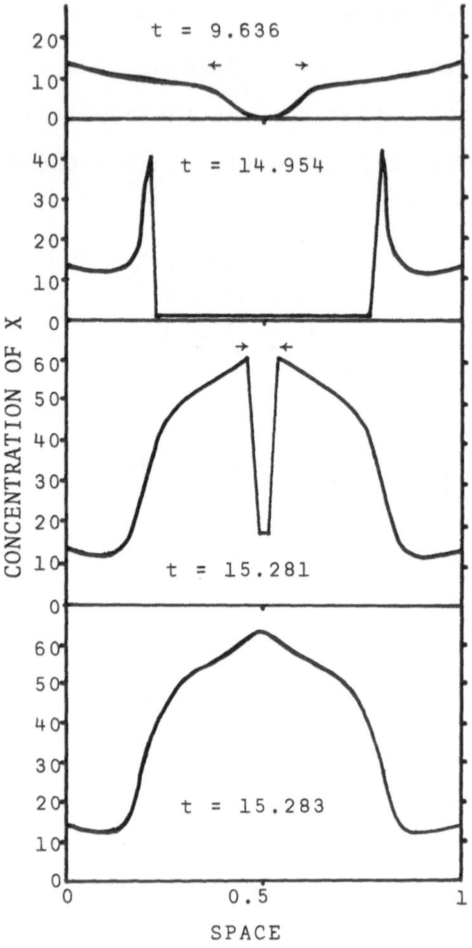

Fig. 8. A typical sequence of profiles in the cycle of a spatio-temporal dissipative structure.

* For further discussion of "chemical evolution" and its biological implications see Ref. [1].

between the two built-up points) the system performs locally discontinuous or relaxation oscillations (e.g., Fig. 9). It is important to note that these nonlinear oscillations do not organize to form a standing wave. Instead there appear (in different parts of a cycle) two distinct types of propagating wavefronts corresponding to inward and outward waves. An interesting consequence of the nonlinear character of the reaction scheme, and hence also of the waves, is the concentration dependence (on X) of the velocity of the waves. Typical propagation velocities range from 0.66 to 0.94, compared with corresponding velocities of order 0.04 for simple diffusion of X. Wavelike propagation (of X) is therefore at least an order of magnitude faster than diffusion, and moreover, may even be directed against a concentration gradient and therefore opposite to any diffusive flow. The comments at the end of the preceding paragraph apply to localized spatio-temporal dissipative structures as well. In addition, one may expect that localization can provide a stabilizing mechanism for dissipative structures with respect to abrupt changes in the chemical environment. Moreover, a characteristic property of localized structures is the greatly enhanced production of a particular substance during a short time and in a limited space (Fig. 9). In this manner regulatory functions for localized processes activated above concentration thresholds may be performed. Finally, regulatory and other types of "information" may well be transmitted in the form of chemical signals (i.e., waves) and stored, as indicated earlier, in stationary localized spatial structures.

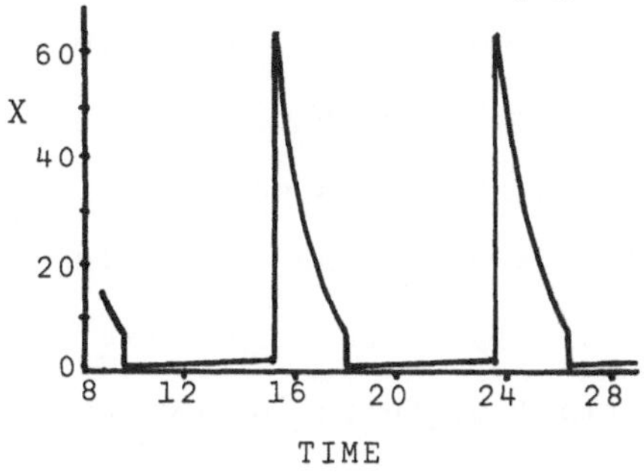

Fig. 9. Variation of X at the middle point (20).

C. Multiple Stationary States.

In this section we focus on systems which may exhibit a number of steady states separated from the "thermodynamic branch" by instabilities which are <u>not symmetry-breaking</u>. Typically the governing evolution equations for such systems may admit two or more spatially homogeneous, time-independent solutions which for a range of constraints are <u>simultaneously</u> accessible to the system. In constrast with the types of dissipative structures discussed earlier, unstable transitions between multiple steady states correspond to a specific <u>functional</u> (rather than temporal and/or spatial) <u>organization</u> of the system. To illustrate the nature of such tansitions we consider a model biochemical scheme* which involves the autocatalytic production of an intermediate compound X and its subsequent enzymatic degradation:

$$
\begin{aligned}
A + X &\underset{k_{-1}}{\overset{k_1}{\rightleftharpoons}} 2X \\
X + E &\underset{k_{-2}}{\overset{k_2}{\rightleftharpoons}} C \\
C &\underset{k_{-3}}{\overset{k_3}{\rightleftharpoons}} E + B.
\end{aligned}
\tag{3.21}
$$

Here A and B represent the reactant and product, E the free enzyme, and C its complex with substrate X. The reservoirs of A and B are time-independent, and in addition the system has access to a fixed total amount of enzyme: $E_T = E + C$. By means of this enzyme conservation condition the macroscopic kinetic equations may be written in the form

$$
\begin{aligned}
\frac{dX}{dt} &= k_1AX - k_{-1}X^2 - k_2XE + k_{-2}(E_T-E) \\
\frac{dE}{dt} &= -k_2XE + (k_{-2}+k_3)(E_T-E) - k_{-3}BE = -\frac{dC}{dt} .
\end{aligned}
\tag{3.22}
$$

At the time-independent regime, the system (3.22) yields, after a few simple manipulations, a cubic equation the solutions of which represent possible concentrations of X at the steady state:

$$
\begin{aligned}
&k_{-1}k_2X_o^3 + [k_{-1}(k_{-2}+k_3+k_{-3}B) - k_1k_2A]X_o^2 + \\
&+ [k_2k_3E_T - k_1A(k_{-2}+k_3+k_{-3}B)]X_o - k_{-2}k_{-3}E_TB = 0
\end{aligned}
\tag{3.23a}
$$

* As in the previous section, this model is chosen more for its analytical simplicity than for any connection with an actual chemical system.

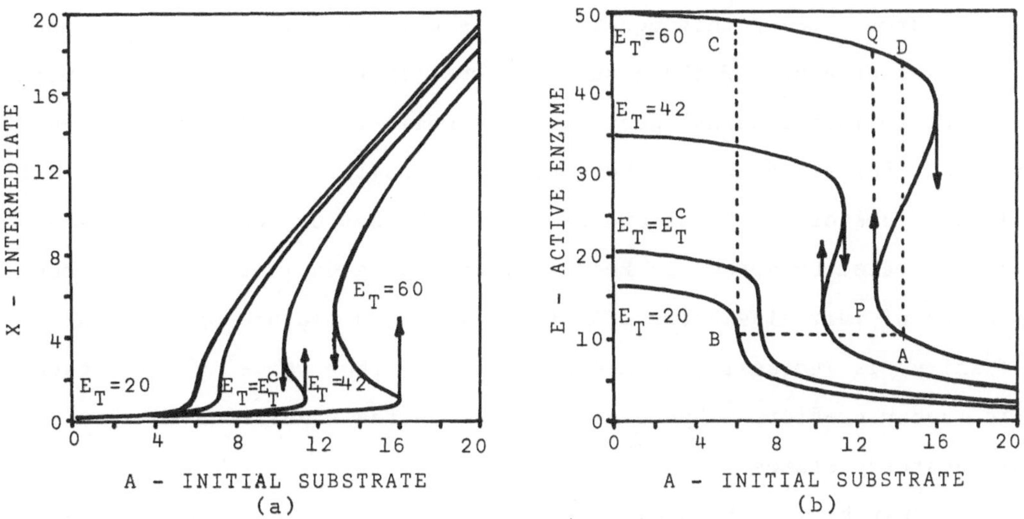

Fig. 10. Steady states of the system (3.21), parameterized by E_T, with B=0.2.

with

$$E_o = \frac{(k_{-2}+k_3)E_T}{(k_{-2}+k_3+k_{-3}B)+k_2X_o}. \tag{3.23b}$$

The conditions for thermodynamic equilibrium are easily seen to be

$$X_{eq} = \frac{k_1}{k_{-1}} A_{eq} = \frac{k_{-2}k_{-3}}{k_2k_3} B_{eq}. \tag{3.24}$$

A typical family of steady state solutions, parameterized by E_T for B=0.2, is displayed in Figs. 10. (For simplicity all rate constants are set to unity.) For E_T less than a "critical" value, $E_T^c \sim 24.8$ in this case, there is a unique solution for all A. Beyond E_T^c, however, there exists a range of A within which appear three possible steady states. A normal mode analysis of Eqs. (3.22) yields the characteristic equation

$$\begin{aligned} \omega^2 &+ [-k_1A+(2k_{-1}+k_2)X_o+k_2E_o+k_{-3}B+k_{-2}+k_3]\omega + \\ &+ (2k_{-1}X_o-k_1A)(k_2X_o+k_{-3}B+k_{-2}+k_3)+k_3E_o(k_{-3}B+k_3)=0, \end{aligned} \tag{3.25}$$

the roots of which determine the stability properties of the alternative states. Using this equation it is easy to verify that the upper and lower branches of the S-shaped curves in Figs. 10 are stable with respect to infinitesimal perturbations while the middle branch is unstable. For fixed $E_T > E_T^c$, therefore, a transition between the two stable branches of the solution occurs at a higher or lower value of A (as indicated by arrows in Figs. 10) depending on whether the concentration of initial substrate A (for fixed B, a measure of chemical affinity) is increasing or decreasing. In other words, the macroscopic theory predicts a hysteresis effect in the transition between multiple steady states.

Just as for the model discussed in Sec. III-B, here the states of marginal stability correspond to vanishing of the coefficients in the characteristic equation (3.25). For $E_T > E_T^c$, we find a pair of nonoscillatory marginal states which are points of transition between stable branches of the steady state curve (arrows, Figs. 10). At $E_T = E_T^c$ the two transition points coalesce into a single "critical" point. In this case both coefficients in Eq. (3.25) vanish, and it is not hard to see (Figs. 10) that the resulting nonoscillatory marginal state corresponds to an inflection point in the steady state curve. Moreover one can verify that the thermodynamic conditions $\delta P = 0 = \delta \Pi$ are fulfilled at the states of marginal stability.

The existence of the critical value E_T^c separating single-valued solutions of Eqs. (3.23) from multi-valued solutions is a remarkable feature not shared by systems exhibiting symmetry-breaking instabilities. Indeed, by exploiting an additional degree of freedom in the system (B or E_T in this example) it is possible to connect any two points on different branches of a multiple steady state curve by <u>continuous</u> paths of steady states. For example, consider a transition from state A to state D of Fig. 10b. The obvious path APQD includes a discontinuous transition at point P. An alternative route involves changing E_T at constant E along AB and at constant A along BC, then varying A at constant E_T along CD. This additional freedom implies that both stable branches of a multiple steady state solution belong to the thermodynamic branch in that each is attainable by continuous extension from a state of thermodynamic equilibrium. (This may be seen as well by varying B rather than E_T.) This striking characteristic of multiple steady states is in sharp constrast with qualitatively new states which appear beyond symmetry-breaking instabilities. States such as stable limit cycles or spatial structures are always separated from the thermodynamic branch by a discontinuous transition originating in an instability due to infinitesimal fluctuations.

From the above discussion we observe that nonequilibrium transitions among multiple steady states have many features in common with equilibrium first-order phase transitions. The van der Waals theory of the liquid-vapor transition, for example, predicts a range of pressure in which equilibrium isotherms (below the critical temperature) exhibit three possible volume states. The middle state is ruled out on physical grounds (negative isothermal compressibility!), while the other two are in principle accessible throughout the coexistence region. The liquid-vapor phase change should occur therefore at a lower pressure than the reverse transition. That a single equilibrium transition pressure is found regardless of the direction of the change is a direct consequence of the response to finite amplitude fluctuations in the initial fluid state.

Pursuing the phase transition analogy further, one would like to know in the present example when the "real" system will actually jump from one branch to another. As the affinity increases, for example (Figs. 10), will the system remain on the initial branch until the point of infinitesimal instability (arrows, Figs. 10), as the macroscopic theory predicts, or will it be driven from that branch at some earlier point in response to finite fluctuations in the medium?

The problem, therefore, for both equilibrium and nonequilibrium transitions, is to understand the range of validity of the purely macroscopic description in the neighborhood of an instability. To approach this question of stability with respect to finite fluctuations for nonequilibrium states we must, as in the equilibrium case, turn to a molecular description of the phenomenon. Unfortunately, a complete microscopic treatment in terms of nonequilibrium statistical mechanics is not yet possible. Nevertheless, a first step in this direction is provided for chemical systems by the adoption of a stochastic approach to chemical kinetics.* Recently a stochastic model for the system treated here has been analyzed [13], the results indicating that in this system the transition will occur at a unique value of A and the hysteresis effect will in fact not appear. A possible thermodynamic interpretation of this startling result, which has been found as well in other systems [14], has also been presented [15].

CHAPTER IV: FLUCTUATIONS

A. Introduction.

In the discussions and illustrations of the preceding two chapters the fundamental importance of fluctuations has been emphasized repeatedly. In general terms, we have seen the emergence of a noneqilibrium concept of order through fluctuations, a new ordering principle which refers to the amplification of fluctuations beyond an instability and to their ultimate stabilization by continuous exchange of matter and/or energy with the surroundings. Inspired by the wealth of new types of behavior induced beyond instability by specific types of fluctuations, and motivated in part by a need to understand more completely the mechanism of macroscopic instabilities, we turn in this chapter to the problem of incorporating into our description of nonlinear systems a consistent treatment of fluctuations.

*The stochastic method is discussed in the next chapter.

In a system characterized by a large number ($N \sim 10^{23}$) of degrees of freedom, fluctuations must be taken into account as soon as one adopts a macroscopic description. In such a description the state of a system is specified by assigning values to a limited number ($n \ll N$) of independent variables (e.g., p, T). Such a macroscopic state is always associated with rapid transitions among different atomic states, however, and therefore will constantly experience instantaneous deviations from specific average values. These deviations (or fluctuations), although mechanical in origin, nevertheless appear to a macroscopic observer as completely random events. Therefore the appearance of a fluctuation of a given type is fundamentally a stochastic process. The response of the system to this fluctuation, on the other hand, will obey purely macroscopic laws as long as the system can damp the fluctuation and return to the reference (i.e., average) state. According to classical (equilibrium) thermodynamics, fluctuations are always small and the latter situation will prevail, except in the vicinity of phase transitions. In Chapter II we saw that nonlinear irreversible thermodynamics also predicts instabilities and transitions for certain types of systems maintained far from equilibrium. Precisely as for equilibrium (phase) transitions, then, near such instabilities fluctuations may be amplified and give rise to observable effects. Whatever the outcome of a macroscopic instability, therefore, the instability itself originated in the response of the system to a fluctuation. Moreover, as was seen in Chapter III, the behavior of the system beyond instability is determined largely by the specific type of fluctuation to which the system is initially unstable. Consequently a purely deterministic description of the system in terms of mean values alone is no longer adequate. In this case it is essential to supplement the "average" thermodynamic description by a theory of fluctuations extended to include (nonlinear) far-from-equilibrium situations.

B. Fluctuations Around Nonequilibrium Steady States.

We begin by considering how the probability of occurrence of a specific type of fluctuation may be determined, and look first at how this problem is solved at thermodynamic equilibrium. In an isolated system, Boltzmann's relation defining entropy in terms of the number of states accessible to a system may be inverted to give the classical Einstein formula for the probability of fluctuations around a macroscopic equilibrium state, $P \sim \exp[\Delta S/k]$, where the entropy change is negative for a fluctuation. For small fluctuations ΔS may be expanded to second order as in Eq. (2.14). Since for an isolated system S is maximum at equilibrium, the probability expression reduces to

$$P \sim \exp[(\delta^2 S)_e/2k], \tag{4.1}$$

where $(\delta^2 S)_e$ is evaluated around the equilibrium state.

In order to discuss fluctuations near macroscopic instabilities far from thermodynamic equilibrium, it is necessary first to generalize Eq. (4.1) to nonequilibrium situations. As was demonstrated in Sec. II-B-3, the basic property $\delta^2 S<0$, which is responsible for the validity of Eq. (4.1), is shared by fluctuations around arbitrary nonequilibrium states for systems in which local equilibrium is maintained. Hence Prigogine suggested some time ago that a nonequilibrium fluctuation theory be founded, in first approximation, on the basic relation (4.1), wherein the excess quantities would be calculated around nonequilibrium reference states:

$$P \sim \exp[(\delta^2 S)_o/2k] \ . \tag{4.2}$$

The justification of such a theory presents formidable difficulties, inasmuch as a microscopic or molecular treatment is essential to any complete proof. In recent years, however, significant advances in this direction have been made by adopting a stochastic description of many-body systems. Before discussing these developments in more detail, we enumerate briefly the principal results:

(a) Within the entire domain of validity of the local equilibrium theory, general arguments and specific model calculations imply that the probability of small thermal fluctuations around a macroscopic reference state is given by the generalized Einstein relation (4.2). As long as $(\delta^2 S)_0 < 0$ and $(\partial/\partial t)(\delta^2 S)_0 > 0$, then, small fluctuations decay and the reference state is asymptotically stable. Once the latter inequality is violated, however, small fluctuations may grow and lead the system to a new regime.

(b) Since dissipative instabilities for systems in local equilibrium are purely macroscopic phenomena having no direct molecular analogs, it is a reasonable expectation that systems undergoing such macroscopic instabilities cannot evolve from a given macroscopic reference state in response to small thermal fluctuations.*

(c) A change in a macroscopic (local equilibrium) state must therefore involve a mechanism of large thermal fluctuations of macroscopic size (consistent with an analogy to first-order phase transitions). The probability of such fluctuations being very small, one expects the Einstein relation (4.2) to be a good approximation even for fluctuations around the time-dependent average of a macroscopic evolution. Hence the time necessary to establish such an evolving mode should also be macroscopic (i.e., long compared to the relaxation time between molecular collisions), and in fact comparable to the "hydrodynamic" scale of macroscopic evolution.

(d) The probability that a finite fluctuation occurs everywhere in a macroscopic system should be negligibly small. One expects, rather, that a small subsystem first begins to evolve in response to a finite fluctuation, creating a local inhomogeneity which would ordinarily be damped by diffusion. If the system is near the threshold of a macroscopic transition, however, diffusion provides as well a mechanism by which an initial disturbance exceeding the threshold can be propagated over macroscopic distances, involving therefore the entire system in a transition which began as a local instability.

---

*Phenomena such as plasma instabilities or laser thresholds are instabilities in the velocity distribution, and therefore cannot be described by a local equilibrium theory. Such "molecular instabilties" appear analogous to second-order phase transitions, while macroscopic instabilties are more closely related to first-order phase transitions.

C. The Markovian Stochastic Approximation.

In view of as yet unresolved difficulties encountered in a microscopic treatment of fluctuations (via nonequilibrium statistical mechanics), we adopt a stochastic approach intermediate between a rigorous microscopic theory and the macroscopic description. Consider the set $\{a_i\}$ of variables necessary to specify the macroscopic state of a system. As has already been pointed out, variations in the a_i due to fluctuations appear as random or stochastic events. One imagines, therefore, an ensemble of replicate systems among which instantaneous values $a_i(t)$ will in general be distributed about mean values $\overline{a_i(t)}$ taken over the ensemble. In a stochastic theory the system is fully described once a hierarchy of probabilities $P_k(a_1,t_1;\, a_2,t_2 \ldots a_k,t_k)da_1 \ldots da_k$, $k = 1, 2\ldots$, of finding a_1 within (a_1, a_1+da_1) at time t_1, a_2 within (a_2, a_2+da_2) at time t_2, etc., has been determined. It is convenient also to introduce conditional probabilities W_k defined by

$$P_1 = W_1 \quad , \quad P_2(a_1,t_1;a_2,t_2) = P_1(a_1,t_1)\, W_2(a_1,t_1|a_2,t_2), \text{ etc.} \qquad (4.3)$$

For a wide variety of problems, W_2 contains all essential information, and therefore is said to represent the transition probability for a Markov process. If in addition the doublet probabilities depend on time only through differences (stationary Markov process), then the treatment of fluctuations reduces to the construction and solution for the problem at hand of an integral equation of the general form

$$P_2(a_1;a_2,t) = \int da\, W_2(a|a_2,s)P_2(a_1;a,t-s) \quad , \quad 0 \leq s < t \ . \qquad (4.4)$$

Taken alone, this equation admits a single time-independent solution corresponding to thermodynamic equilibrium. For small fluctuations, moreover, the classical Einstein formula (4.1) is recovered. Far from equilibrium, the solution of Eq. (4.4) is subject to the same constraints which are responsible for maintaining a steady nonequilibrium state, and the problem of fluctuations is no longer tractable in its most general form. We turn, therefore, to model calculations which illustrate the general approach, and consider for simplicity a sequence of two monomolecular chemical reactions:*

$$A \underset{k_{21}}{\overset{k_{12}}{\rightleftarrows}} X \underset{k_{32}}{\overset{k_{23}}{\rightleftarrows}} E. \qquad (4.5)$$

The affinity of the overall reaction $A \rightleftarrows E$ is $A = RT \ln(KA/E)$ (where $K = k_{12}k_{23}/k_{21}k_{32}$ is the equilibrium constant), and can be assigned arbitrary values depending on the concentrations of A and E. If the concentrations A and E are fixed, then for an ideal solution the single macroscopic kinetic equation which determines the evolution of the (spatially uniform) system is

$$\frac{dX}{dt} = (k_{12}A + k_{32}E) - (k_{21} + k_{23})X , \qquad (4.6)$$

which has a unique steady state solution

$$X_o = \frac{k_{12}A + k_{32}E}{k_{21} + k_{23}} . \qquad (4.7)$$

In the deterministic approach X represents the <u>average</u> concentration of constituent X, etc., and fluctuations about such average values are neglected. To discuss fluctuations, therefore, we require a more refined description. For a system undergoing chemical reactions the simplest possible stochastic model assumes that the mass-conservation equations (4.6) [c.f., Eqs. (2.2)] define a Markov process in the <u>number-of-particles space</u>.** Accordingly we denote the <u>numbers</u> of molecules of type A,X,E by the <u>discrete</u> random variables a(t), x(t), e(t). The state of the system is then prescribed by the probability distribu-

---

*For chemical reactions without transport the stochastic problem may be formulated in terms of a <u>finite</u> number of <u>discrete</u> variables (e.g., the number of particles of various species), and Eq. (4.4) has then a comparatively simple form.

**One expects that a stochastic model for chemical kinetics should be valid for reactions which are not <u>too</u> fast. For a discussion of theory and application of the "birth-and-death" stochastic approach used here, see the review by McQuarrie [16].

tion P(a,x,e;t) of those random variables. An evolution equation for this function is constructed by considering separately the (additive) contributions for the four processes.

For the first reaction $A \xrightarrow{k_{12}} X$, we have the contribution

$$P(a,x,e;t+\Delta t)=W_{12}(a+1)P(a+1,x-1,e;t) + [1-W_{12}(a)]P(a,x,e;t) \quad (4.8)$$

corresponding to the probability of conversion $(a+1,x-1) \longrightarrow (a,x)$ plus the probability of no conversion $(a,x) \longrightarrow (a-1,x+1)$. In an ideal (dilute) solution the transition probability W_{12} may be written

$$W_{12}(a)=\ell_{12}a\Delta t+\Theta(\Delta t^2), \quad (4.9)$$

where ℓ_{12} is a constant (related to k_{12}) and Δt is a time interval which is small enough that only one molecule A will undergo a transition. Considering now all four reactions, and taking the limit $\Delta t \dashrightarrow 0$, we obtain a finite difference equation with linear coefficients (linearity is due to monomolecular character of the reaction):

$$\begin{aligned}\frac{dP(a,x,e;t)}{dt} &= \ell_{12}(a+1)P(a+1,x-1,e;t) - \ell_{12}aP(a,x,e;t)\\ &+ \ell_{21}(x+1)P(a-1,x+1,e;t) - \ell_{21}xP(a,x,e;t)\\ &+ \ell_{23}(x+1)P(a,x+1,e-1;t) - \ell_{23}xP(a,x,e;t)\\ &+ \ell_{32}(e+1)P(a,x-1,e+1;t) - \ell_{32}eP(a,x,e;t).\end{aligned} \quad (4.10)$$

Equations of this type are most conveniently studied in the generating function representation. We define the generating function*

$$F(s_a,s_x,s_e;t)\equiv\sum_{a,x,e} s_a^a s_x^x s_e^e P(a,x,e;t), |s_i|\le 1 (i=a,x,e), \quad (4.11)$$

which has the following properties (s=1 implies $s_a=s_x=s_c=1$):

$$(F)_{s=1}=1 \quad , \quad \left(\frac{\partial F}{\partial s_x}\right)_{s=1} = \sum_{a,x,e} xP = \bar{x}$$

$$\left(\frac{\partial^2 F}{\partial s_x^2}\right)_{s=1} = \sum_{a,x,e} x(x-1)P=\overline{x^2}-\bar{x}, \text{ etc.} \quad (4.12)$$

*Sums over population variables are from zero to infinity.

Introducing Eq. (4.11) into Eq. (4.10) yields a first-order partial differential equation with linear coefficients:

$$\frac{\partial F}{\partial t}=\ell_{12}(s_x-s_a)\frac{\partial F}{\partial s_a}+\ell_{21}(s_a-s_x)\frac{\partial F}{\partial s_x}+ +\ell_{23}(s_e-s_x)\frac{\partial F}{\partial s_x}+\ell_{32}(s_x-s_e)\frac{\partial F}{\partial s_e} \quad . \tag{4.13}$$

The only asymptotic solution to this equation corresponds to thermodynamic equilibrium. In order to introduce the constraints (fixed A,E) which permit a stationary nonequilibrium solution of Eq. (4.13), we consider the reduced probability distribution

$$\rho(x;t)=\sum_{a,e}P(a,x,e;t)$$

and (4.14)

$$f(s;t)=\sum_{a,x,e}s^{x}P(a,x,e;t)=F(s_a=1,s_e=1,s_x=s;t).$$

Instead of Eq. (4.13) we now have, from Eq. (4.10),

$$\frac{\partial f}{\partial t}=(1-s)[(\ell_{21}+\ell_{23})\frac{\partial f}{\partial s}-(\ell_{12}A+\ell_{32}E)f], \tag{4.15}$$

where it has been assumed that the state of reservoirs A and E is not influenced by the internal state of the system.* Mathematically this implies, for example,

$$\sum_{a,e}aP=\bar{a}\sum_{a,e}P=A\rho(x;t), \tag{4.16}$$

where the identifications $A\equiv\bar{a}$, $E\equiv\bar{e}$, introduce as a parameters in the theory the boundary values of A,E.

Solution of Eq. (4.15) yields a unique steady state

$$f(s)=\exp[(s-1)\frac{\ell_{12}A+\ell_{32}E}{\ell_{21}+\ell_{23}}] \tag{4.17}$$

which predicts a steady state average value $X_o=\bar{x}_o=(\partial f/\partial s)_{s=1}$ identical

---

*This decoupling procedure implies a wide separation of time scales characteristic of composition changes into a large one for the reservoir variables (A,E) and a smaller one for the internal system variables (X). This condition is nothing more than the approximation made in defining what is meant by a steady nonequilibrium state.

with Eq. (4.7) provided the identification $\ell_{ij}=k_{ij}(i\neq j=1,2,3)$ is made. Thus

$$f(s)=\exp[(s-1)X_o], \tag{4.18}$$

and in physical variables the solution is a Poisson distribution

$$\rho(x)=e^{-X_o}\, X_o^x/x! \tag{4.19}$$

If small fluctuations $\delta x=x-\bar{x}$ are considered, then it is easy to show that Eq. (4.19) reduces to a Gaussian distribution:

$$\rho(x)=(2\pi\bar{x})^{-\frac{1}{2}}\exp(-\delta x^2/2\bar{x}), |\delta x/\bar{x}|<<1 \quad . \tag{4.20}$$

Recalling now Eq. (2.15), and noting that $\partial\mu_i/\partial\bar{N}_j=kT\delta_{ij}/\bar{N}_i$ for an ideal chemical mixture, we find that the second-order excess entropy becomes $\delta^2S=-k\delta x^2/\bar{x}$. Therefore the distribution of small fluctuations (4.20) is precisely the generalized Einstein formula (4.2) for the model (4.5).

The results of this analysis have been extended to more complicated systems [17]. The conclusion is that in the most general sequence of monomolecular reactions in ideal mixtures arbitrarily far from equilibrium, the probability of fluctuations around the steady state population of an intermediate species is given by a Poisson distribution, and reduces for small fluctuations to the form (4.20) [and therefore also (4.2)]. For a number of models involving bimolecular reactions, however, the birth-and-death stochastic analysis yields distributions which are not of the Poisson type [18]. Moreover, the differences from a Poisson distribution are found to depend on specific properties of the various bimolecular steps. It would appear, therefore, that the generalized Einstein relation applies out of equilibrium only to monomolecular reactions (and of course to arbitrary systems at equilibrium). Before drawing any definite conclusion, however, we take a closer look at the assumptions underlying the thermodynamic and stochastic descriptions.

D. The Need for a More Complete Description.

Within the framework of nonlinear thermodynamics, the quantity $\delta^2 S$ was seen in Chapter II to provide a measure of fluctuations around a macroscopic reference state. In addition, for equilibrium systems and for nonequilibrium systems of monomolecular reactions, the connection between the "entropy of fluctuations" $\delta^2 S$ and actual statistical properties of fluctuations is made quantitative by a formula of the Einstein type (4.1) or (4.2). An important result, therefore, is that for such systems, at least, the theory of fluctuations around nonequilibrium states can be formulated in terms of precisely the same thermodynamic functions (defined locally) as at equilibrium. The underlying foundation for such a conclusion is implicit in the assumption of local equilibrium on which the extension of thermodynamics to nonequilibrium situations is based. Why then should one not expect the generalized Einstein relation to remain valid whenever the local equilibrium assumption is satisfied?

In the current context local equilibrium implies that the velocity distribution for each component of a chemical mixture deviates little from a local Maxwellian. On a molecular level continuous restoration to a local Maxwellian is accomplished by frequent elastic collisions (with inert solvent molecules) between relatively infrequent inelastic (reactive) collisions which tend to distort the velocity distribution. Hence for local equilibrium to be guaranteed for <u>all</u> situations, a theory of fluctuations in chemically reacting mixtures must take explicitly into account the existence of two widely separated time scales characterizing these two competing collisional processes.

It is clear from preceding sections that a stochastic theory based on a birth-and-death formulation of chemical kinetics cannot ensure the maintenance of a local equilibrium regime [18]. On the contrary, the last paragraph suggests that a proper description of fluctuations requires the details of the behavior of the system on a molecular level, as given, for example, by a microscopic kinetic equation of the Boltzman type. Assuming that such an equation defines a Markov process in the complete phase space, including internal states, one obtains a generalized stochastic master equation in that space. By solving such an equation for specific nonlinear, nonequilibrium, chemical models, Nicolis and Prigogine [18] show that indeed small fluctuations behave in agreement with the generalized Einstein formula, provided the system can be treated as a dilute (ideal) mixture. Concluding therefore that in nonlinear systems the usual birth-and-death type of stochastic approach gives an inadequate picture of fluctuations and must be replaced by a phase-space description, these authors suggest that throughout the domain of validity of the local equilibrium assumption the generalized Einstein formula describes correctly the behavior of small fluctuations around macroscopic reference states.

E. Finite Fluctuations.

So far we have been dealing with fluctuations around states which are asymptotically stable, and which moreover are far removed from unstable transition points. For a system describable by a local equilibrium theory, the stability properties are expressed in terms of the excess entropy. The probability of small fluctuations around a steady state is given then by the generalized Einstein relation (4.2), which implies, since $(\delta S)_0 < 0$ [Eq. (2.15)], that this state is more probable than all neighboring states accessible by small fluctuations. Suppose the constraints on the system change until the state becomes unstable. Just before the point of marginal stability, Eq. (4.2) still predicts that the reference state is most probable. Once inequality (2.16) is violated, however, small fluctuations will be amplified to finite size, driving the system to a new regime around which fluctuation will again be small and described by Eq. (4.2). A useful schematic of this situation appears in Fig. 11.

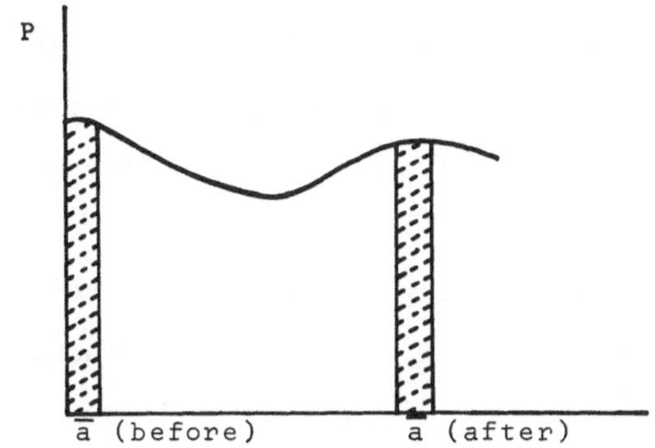

Fig. 11. The probability function in terms of a fluctuating variable a. Dashed areas denote domains of validity of the generalized Einstein Eq. (4.2).

Evaluation of the probability function for nonlinear systems undergoing macroscopic instabilities is a difficult problem. On the other hand, the Volterra-Lotka model

$$\begin{aligned} A + X &\rightleftharpoons 2X \\ X + Y &\rightleftharpoons 2Y \\ Y + B &\rightleftharpoons E \end{aligned} \tag{4.21}$$

has been found to behave like a system which is permanently at a marginal state [8]. In this model, the single steady state is surrounded by an infinity of closed orbits. Neither the steady state nor the orbits, however, is asymptotically stable, there being no mechanism in this model for regression of fluctuations. Hence one expects the study of fluctuations in this model to yield valuable information regarding the behavior of systems undergoing unstable transitions.

The Volterra-Lotka model has been analyzed in detail by Nicolis and Prigogine [8,18]. Of particular importance here is the conclusion from a phase-space description [18] that the behavior of small fluctuations around the steady state is still described by the generalized Einstein formula (4.2). Evolution from that state, and indeed among any of the available orbits, is governed by the appearance of finite amplitude fluctuations in the system. The probability of occurrence of such fluctuations of arbitrary size is obviously not expected to have the form of a generalized Einstein distribution.

Also in their analysis, the inadequacy of the birth-and-death formulation of chemical kinetics is quite transparent. Indeed, within such a description, the first step of scheme (4.21) implies that two molecules of type X are created in precisely the same internal state as the X molecule which combines with A. In a very large system, however, the probability of this event should be negligible compared to the same process in which now each X molecule may be in a distinct internal state. Consequently a stochastic master equation of the birth-and-death type cannot describe correctly the fluctuations in a thermodynamic system.

In summary, we have seen that the decoupling of macroscopic behavior from fluctuations, which is guaranteed at or near thermodynamic equilibrium (except near phase transition or critical points), is no longer possible in general far from equilibrium, and is especially doubtful for systems which may experience macroscopic instabilities. In such cases it becomes essential to study simultaneously, in a self-consistent fashion, evolution of the fluctuations along with evolution of the macroscopic state. Specifically, this implies that a phase-space description, including internal states, must be adopted. This approach in general presents a number of difficulties, especially for nonlinear systems, and is at present the subject of vigorous research.

PART TWO

APPLICATION TO BIOLOGY: THE CONTROL OF METABOLIC AND BIOSYNTHETIC PROCESSES

INTRODUCTION

We have seen in PART ONE that open systems undergoing nonlinear chemical reactions and diffusion far from thermodynamic equilibrium are subject to instabilities in the "thermodynamic branch" of solutions and may evolve subsequently to other operating regimes characterized by new types of behavior. The class of systems which may exhibit such behavior is characterized by two essential features. On one hand, certain types of nonlinearities are required both to permit the existence of more than one steady state and to provide an appropriate chemical mechanism by which instabilities may appear. On the other hand, it is necessary that the system operate in the far-from-equilibrium (nonlinear) domain of irreversible thermodynamics. That is, a minimum level of dissipation is essential to both creation and maintenance of the "dissipative structures" which may arise beyond instability. In biological systems there exists an essentially limitless variety of mechanisms and pathways which satisfy both requirements. Two such mechanisms are presented in the following chapters, followed by an Appendix containing references to other examples.

The necessity in biology for elaborate control mechanisms which ensure that the various chemical reactions in living cells happen at the proper rate and at the right time is well known. Actually, one may distinguish between two types of control systems. The first type makes sure that there is neither excessive nor insufficient synthesis of small metabolites, e.g., energy-rich molecules such as ATP. The usual way in which this mechanism operates is to affect the rate at which a particular protein (enzyme) catalyzing a single reaction step acts. The second type of control mechanism affects the rate of synthesis of the various protein molecules in the cell. We shall now discuss one representative example for each of the two types of mechanism.

CHAPTER V: TRANSFER OF CHEMICAL ENERGY IN LIVING CELLS--GLYCOLYSIS

A. Introduction. [19]

The living cell is a unit which maintains an incredibly complex spatial and temporal order only by constant exchange of energy and matter with its environment. Its basic energy requirements are two-fold: In addition to the chemical work necessary to preserve the integrity of its organization, individual cells must transform energy to do various kinds of mechanical, electrical, chemical, and osmotic work that constitute the life processes of living organisms. The living cell is under certain constraints in obtaining and using this required energy: It must operate at relatively low temperature within a narrow range, in an aqueous environment, and within narrow bounds on hydrogen ion concentration. To secure its primary energy, therefore, the cell has evolved molecular mechanisms which work with great efficiency under these mild conditions.

All living cells ultimately derive their energy from sunlight--plant cells directly via photosynthesis, and animal cells indirectly via oxidation of foodstuff molecules (respiration). Whichever primary energy-extracting mechanism in employed, the same molecule--adenosine triphosphate (ATP)--carries the free energy obtained from foodstuffs or from sunlight to all energy-expending processes of the cell. In the cell, the terminal phosphate group of the three linked phosphate groups of ATP is detached by hydrolysis <u>and</u> transferred to a specific acceptor molecule. The free energy of the ATP molecule (associated with the high-energy terminal phosphate bond) is largely conserved in these coupled processes by phosphorylation of the acceptor molecule, the free energy of which is now increased so that it can participate in an energy-requiring process such as biosynthesis or muscle-contraction. Left over from this procedure is adenosine diphosphate (ADP), an energy-poor form of the energy carrier which subsequently may be "recharged" to the energy-rich form by one of the two energy-extracting mechanisms.

The end result of photosynthesis is the production from carbon dioxide and water of glucose, a primary foodstuff molecule from which the energy of molecular configuration may be extracted and conserved in the phosphate-bond energy of ATP. In the cell the recovery of energy from glucose proceeds in two major phases. In the first phase, glycolysis, the glucose molecule is split into two three-carbon molecules of lactic acid, with a net gain of two molecules of ATP. Under anaerobic conditions (i.e., no oxygen), the cell must function on this relatively small (3%) yield of the total energy bound in glucose. In aerobic cells, however, a major portion of the remaining energy is extracted via respiration. In this oxidation process, the three-carbon molecules of lactic acid are broken down to single-carbon molecules of carbon dioxide, with the net production of 36 additional molecules of ATP. Incredibly, the efficiency of energy recovery for the combined processes of glycolysis and respiration is at least 55%! Clearly the need of living cells for highly efficient energy conversion under restricted conditions accounts largely for the enormous complexity of the enzymatic reaction chains responsible for the chemically simple task of converting glucose to carbon dioxide and water.

Having now a general picture of the cellular energy cycle, we turn to a more detailed study of that phase of the energy-recovery process which is common to both aerobic and anaerobic systems--glycolysis. In particular, it is the type of glucose breakdown that occurs in higher animal species as well as in many microorganisms. The apparently simple process of splitting glucose into two molecules of lactic acid occurs in eleven steps, each catalyzed by a specific enzyme. Each of the intermediate products contains phosphate groups, and a net conversion of two molecules of ADP (plus phosphate) to ATP is the overall result (Fig. 12).

B. The Phosphofructokinase System.

We see in Fig. 12 that in some of the eleven sequential reaction

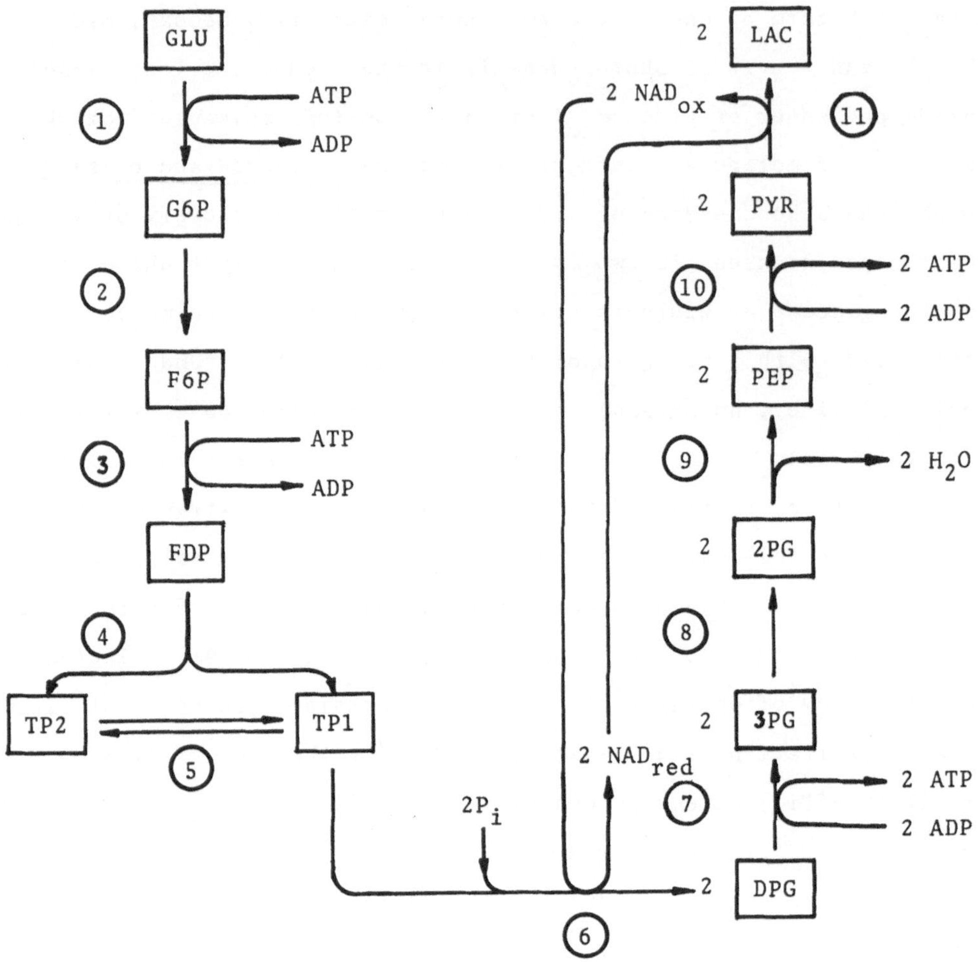

Fig. 12. The sequential reactions of glycolysis, each catalyzed by a specific enzyme. P_i is inorganic phosphate, NAD is nicotinamide adenine dinucleotide.

ENZYMES

1. Hexokinase
2. Phosphohexoisomerase
3. Phosphofructokinase
4. Aldolase
5. Triose isomerase
6. Glyceraldehyde 3-phosphate dehydrogenase
7. 3-phosphoglycerate kinase
8. Phosphoglyceromutase
9. Enolase
10. Pyruvate kinase
11. Lactate dehydrogenase

INTERMEDIATES

GLU	Glucose
G6P	Glucose 6-phosphate
F6P	Fructose 6-phosphate
FDP	Fructose 1,6-diphosphate
TP2	Dihydroxyacetone phosphate
TP1	Glyceraldehyde 3-phosphate
DPG	1,3-diphosphoglycerate
3PG	3-phosphoglycerate
2PG	2-phosphoglycerate
PEP	Phosphoenolpyruvate
PYR	Pyruvate
LAC	Lactate

steps of glycolysis, ATP and ADP are absorbed from or released into the medium. Suppose now that ready-made ATP is supplied to the glycolytic chain from elsewhere. Then, if the enzymes catalyzing the steps go on acting at full speed, too much ATP would be produced. Thus it may be important for the cell to be able to alter (decrease in this case) the catalytic activity of some of the enzymes and slow down the production of ATP. An ingenious feature of the glycolytic chain (and also of other biochemical pathways) is that this control need not act on all the enzymes of the chain, but usually acts only on one of the first. If the action of the first enzyme is decreased, for example, then the action of subsequent enzymes is necessarily decreased also because of substrate limitation.

For a wide variety of biological mechanisms, this type of control may appear in the form of well-defined oscillations in the concentrations of various substrates, products, etc., associated with the mechanism. The most completely investigated and best understood example of such behavior at the metabolic level occurs in glycolysis [20]. Oscillations of NADH were first reported by Duysens and Amesz [21]. Subsequently Chance and coworkers have reported damped oscillations in NADH in suspensions of intact yeast cells [22] plus damped <u>and</u> sustained oscillations in cell-free extracts [23]. Glycolytic oscillations have also been observed in suspensions of tumor cells [24] and in beef-heart extracts [25]. Indeed, eventually <u>all</u> glycolytic metabolites [23], as well as CO_2 production and pH [26,27], have been found to oscillate.

Extensive experimental investigations in yeast have revealed that the most important level of control in glycolysis lies in the action of the enzyme phosphofructokinase (PFK) [20,27,28]. This enzyme is <u>allosteric</u>, and thus has remarkable regulatory properties on the molecular level. Each subunit of such an allosteric molecule (PFK is a tetramer) has two binding sites. One site on each unit of PFK binds one of the products ADP, FDP or a substrate F6P, which act as activators. In this configuration, the enzyme is active. The other site of each subunit binds the substrate ATP. In this configuration the enzyme is inactive since the properties of the first site are altered in such a way that binding of activator molecules is precluded. Thus the activity of the enzyme PFK depends on the metabolite concentrations. Oscillatory or other regulatory behavior is then propagated throughout the glycolytic reaction chain by the coupling of the phosphofructokinase system with all control points along the pathway. We have therefore a possibility for regulating automatically the velocities of production according to the flow of metabolites, which itself depends on the level of metabolite concentration in the cell.

How is such regulation manifested? We have already noted the observation of sustained glycolytic self-oscillations having perfectly reproducible periods and amplitudes. What is then the role of this oscillatory control, and what are the basic reasons for it? We shall postpone an answer to the first question until the next section. Here we shall try to relate these oscillations to the molecular structure of phosphofructokinase as well as to the properties of the PFK scheme of reactions.

The simplest kinetic model of the PFK reaction that one can imagine should represent a single substrate S (ATP) and a single product P (ADP) in a reaction scheme the enzyme of which is activated by the product. Mathematical models for this type of process have been worked out by Sel'kov [29,30] and by Higgins [31]. Following Sel'kov, we express this scheme as follows:

(a) Substrate enters irreversibly at rate v_1.

$$\xrightarrow{v_1} S \tag{5.1a}$$

(b) Substrate combines with E, the <u>active</u> form of phosphofructokinase. Chemical studies show that this reaction proceeds via the formation of an intermediate complex which then decomposes to give the product. The original enzyme is then recovered.

$$S + E \underset{k_{-1}}{\overset{k_{+1}}{\rightleftharpoons}} C \xrightarrow{k_{+2}} E + P \tag{5.1b}$$

(c) Product diffuses irreversibly out of the system.

$$P \xrightarrow{v_2} \tag{5.1c}$$

(d) We must now express the regulatory action of E. This is a fairly complicated matter which will be discussed more fully later. Here we adopt a phenomenological approach and write the global reaction:

$$\gamma P + E^* \underset{k_{-3}}{\overset{k_{+3}}{\rightleftharpoons}} E \quad , \tag{5.1d}$$

where E* is the inactive form of the enzyme which is activated upon binding γ product molecules.

We observe immediately that, due to the last step (i.e., due ultimately to the properties of PFK), the system is nonlinear. Furthermore, because of steps (a) and (c) it is open and operates far from equilibrium. We shall show [29] that these elements will be sufficient to explain the experimental data. That this is so should come as no surprise, since we have seen in PART ONE that nonlinear chemical systems may indeed exhibit oscillations under highly nonequilibrium conditions. For notational convenience we set $s \equiv [S]$, $p \equiv [P]$, $e \equiv [E^*]$, $x_1 \equiv [E]$, $x_2 \equiv [C]$. In the following analysis we assume for simplicity that the system is maintained uniform. In addition we assume that the following conditions are met:

$$\text{(A1)} \quad \gamma > 1 \tag{5.2}$$

$$\text{(A2)} \quad \frac{k_{+1}}{s} , \; k_{-1} , \; k_{+2} , \; \frac{k_{+3}}{p^{\gamma}} , \; k_{-3} >> 1 \tag{5.3}$$

$$\text{(A3)} \quad \frac{s}{e_o} , \; \frac{p}{e_o} >> 1 \tag{5.4}$$

where $e_o = e + x_1 + x_2$ is the total enzyme concentration. Condition (A1) will be discussed later. Condition (A2) is commonly satisfied for most enzyme reactions. Condition (A3) reflects an experimental fact that the substrate and product concentrations in the PFK reaction are much greater (of order 1mM) [32] than the total enzyme concentration (of order 10μM, 1μM, or even less) [32].

A final assumption arises from the fact that the rate of the glycolytic flux in a self-oscillatory state is considerably lower than the maximum rates of the reactions controlling the sink of the products of the PFK reaction [32]. Thus it is appropriate to write a first-order rate law for the product sink:

$$\text{(A4)} \quad v_2 = k_2 p \quad . \tag{5.5}$$

According to the laws of mass action and mass conservation the reaction scheme (5.1) may be described by the following system of rate equations:

$$\frac{ds}{dt} = v_1 - k_{+1} s x_1 + k_{-1} x_2 \tag{5.6a}$$

$$\frac{dp}{dt} = k_{+2} x_2 - k_{+3} p^{\gamma} e \tag{5.6b}$$

$$\frac{dx_1}{dt} = -k_{+1} s x_1 + (k_{-1}+k_{+2}) x_2 + k_{+3} p^{\gamma} e - k_{-3} x_1 \tag{5.6c}$$

$$\frac{dx_2}{dt} = k_{+1} s x_1 - (k_{-1}+k_{+2}) x_2 \tag{5.6d}$$

$$\frac{de}{dt} = -k_{+3} p^{\gamma} e + k_{-3} x_1 \tag{5.6e}$$

By means of assumptions (A2) and (A3) it is possible to invoke a steady state approximation for the transformation (5.1b), while the enzyme activation reaction (5.1d) is itself essentially at equilibrium. After a little algebra the scheme (5.6) reduces then to a pair of coupled equations of the form

$$\frac{ds}{d\tau} = \nu_1 - \nu(s,p) \tag{5.7a}$$

$$\frac{dp}{d\tau} = \alpha[\nu(s,p) - \beta p] \quad . \tag{5.7b}$$

Here the symbols s and p denote dimensionless substrate and product concentrations, τ the dimensionless time variable, α the ratio of the Michaelis constant for substrate to the activation constant for product, ν_1 the (dimensionless) rate of substrate entry and β the corresponding rate of product consumption. Activation by the product is reflected in the fact that $\nu(s,p)$ is an increasing function of p.

It is not difficult to show that the mathematical model (5.7) possesses a single homogeneous solution for a region of parameter values corresponding to experimental situations [29]. For an appropriate range of values of the parameter γ, moreover, that solution is found to be unstable with respect to evolution to a stable limit cycle. Outside that range damped oscillations are predicted as the steady uniform situation is there the stable solution.

We see then that Sel'kov's simple model does in fact exhibit some of the experimentally observed features of the phosphofructokinase system. On the other hand this model is phenomenological in the sense that the parameter γ expresses in a global fashion the properties of the allosteric enzyme. Indeed, as Sel'kov himself points out (assumption A1), for the mathematical model (5.1) to take account of product activation it is necessary that γ be greater than one. The kinetics of the PFK reaction, however, is known to be first order (γ=1). Hence in this model a purely kinetic interpretation of γ as a stoichiometric number of product activation must be abandoned in favor of a mathematical understanding of γ as a measure of the degree <u>both</u> of product activation <u>and</u> of ATP-enzyme interaction.

In order to take into account explicitly the allosteric character of the enzyme, we now turn to a more realistic model of the PFK system presented recently by Goldbeter and Lefever [33]. In this model the activation by the product, the role of the substrate, and the cooperativity are described by three independent parameters which are directly related to experiment. Moreover, the model is sufficiently general to be applied as well to other reaction schemes catalyzed by allosteric enzymes. Here we shall concentrate on the problem of glycolytic oscillations and for simplicity we work with an allosteric dimer [34].

Fig. 13 gives a schematic representation of the model. The main steps are as follows:

(a) The substrate enters the system at a constant rate ν_1;

(b) The enzyme is a dimer which can exist at equilibrium in the form R (active) and T (inactive);

(c) The substrate may react with both forms whereas the product can only bind with the species R;

(d) The various bound enzyme-substrate species decompose irreversibly to give a product;

(e) The product leaves the system at a rate proportional to its concentration.

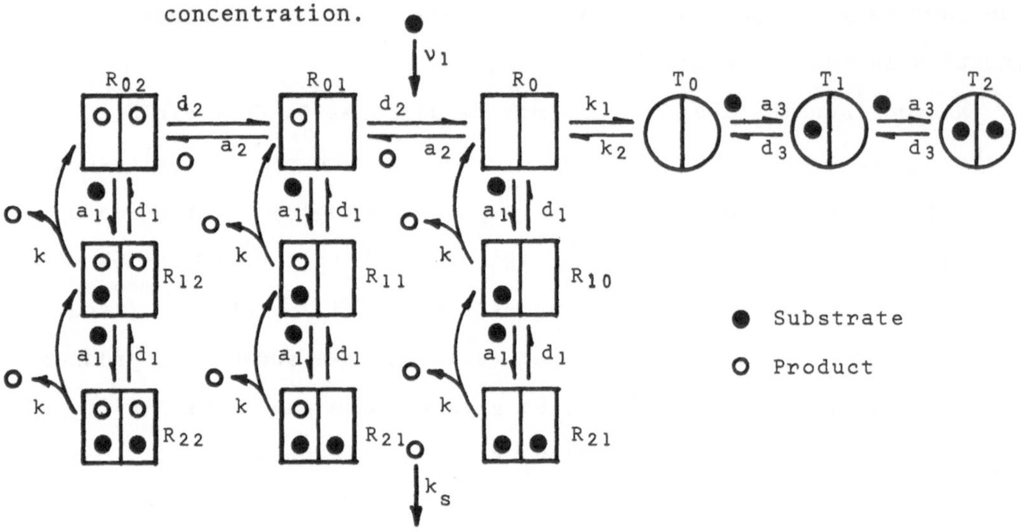

Fig. 13. (after Ref. [33]) Schematic representation of the allosteric model involving activation by the product.

ν_1: rate of entry of the substrate.

R,T: respectively active and inactive forms of the enzyme.

k_1, k_2: kinetic constants for the $R_0 \rightleftharpoons T_0$ reaction.

a_i, d_i: kinetic constants for the reactions involving the binding of the substrate on an enzymatic species.

k: kinetic constant for the (irreversible) decomposition of an enzymatic intermediate into an enzyme and a product.

k_s: kinetic constant for the (irreversible) decomposition of the product.

We observe that steps (a), (d), and (e) are irreversible; this introduces into the problem the nonequilibrium properties which will be responsible for the occurrence of oscillations. Starting from these assumptions and introducing the same type of simplifications (based on experimental data) as in Sel'kov's model, one obtains a set of two coupled equations for the evolution of the substrate and product concentrations. These equations admit a single physical uniform steady state solution which becomes unstable provided a set of conditions between the parameters is satisfied. In particular, the instability condition is fulfilled for large values of the allosteric constant [35] which agree with experimentally observed values. It is also observed that the domain of instability is much wider if the substrate activates the enzyme. In the presence of substrate inhibition instabilities are still possible but disappear beyond a threshold value of the degree of inhibition.

A study of the evolution of the system beyond instability has shown that the system tends to a new stable regime in which the concentrations of the chemicals exhibit sustained oscillations. Fig. 14 shows the variation in time of the substrate and product concentrations for glycolysis [33]. It is seen that when the parameters correspond to experimentally known values the predicted period of oscillation is of the order of a few minutes and agrees therefore with the experimental data [27,28]. For other ranges of values of the parameters the calculation predicts a spatio-temporal dissipative structure in the form of propagating concentration waves [36]. Preliminary steps have already been undertaken to verify experimentally this result [37].

C. The Role of Oscillations in the Control of Biochemical Reactions.

The results of the previous section establish that glycolysis is a chemical clock of the limit cycle type. Thus the observed oscillations can only arise beyond a point of instability of the thermodynamic branch. In other terms, the glycolytic clock is a temporal dissipative structure. In fact, one is tempted to conclude that the stability and uniqueness of most biological rythmic phenomena can only be understood in terms of a limit cycle behavior. This conclusion is not only interesting because it is satisfactory conceptually. By working out the properties of the limit cycle, one can arrive at definite predictions, e.g., calculation of the fraction of time spent by the enzyme in its

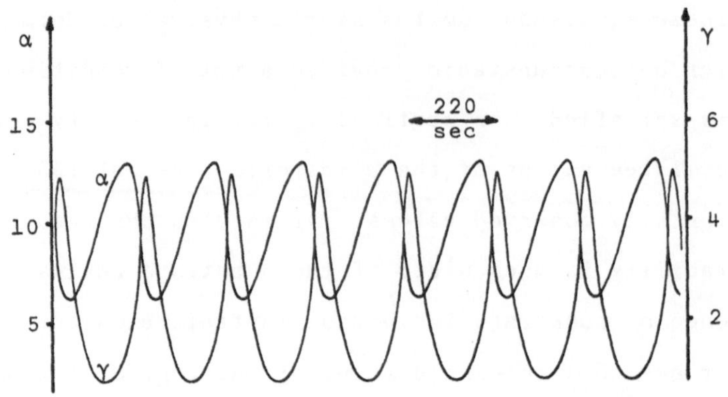

Fig. 14. (after Ref. [33]) Time variation of the ATP and ADP concentrations for glycolysis

$$\alpha = [ATP]\frac{a_1}{d_1} \qquad \gamma = [ADP]\frac{a_2}{d_2}$$

The following numerical values of the parameters have been chosen:

$$L = \frac{T_o}{R_o} = 7.5 \times 10^6 \qquad \text{(allosteric constant)}$$

$$a_i = a = 10^7/\text{mM sec}$$

$$d_i = d = 5 \times 10^5/\text{sec}$$

$$\sigma_1 = \nu_1/d = 10^{-8}\text{mM}$$

$$\sigma_2 = k_s/a = 5 \times 10^{-9}\text{mM}$$

$$\varepsilon = k/d = 10^{-1}$$

$$D_0 = 5 \times 10^{-4}\text{mM} \qquad \text{(total enzyme concentration)}$$

active or inactive form, etc. By comparison with experiment one can obtain in this way indirect information about any uncertain rate constants or other parameters involved in the various reaction steps.

In the particular case of glycolysis one finds that during one period of oscillation the enzymes spend by far most of their time in the inactive form. A similar conclusion has been reached by Walter [38] in analysis of a model biochemical chain involving a step which is inhibited by the final product. Under certain conditions he finds oscillations of the limit cycle type such that the biochemical variables spend considerably more time slightly below the steady state value than in the vicinity above it. One is tempted to think that the biological reason for this behavior is to provide a precisely timed "spike" function of the enzyme or metabolites. In certain limiting cases oscillations can become so sharp that they remind one of all-or-none (flip-flop) type of behavior: the system then exhibits essentially a rhythmic binary "logic".

There are some examples in which the advantage of a "spike" function is obvious. For instance, in certain phenomena, such as mitosis, which are related to overall cell behavior, it is vital that the cell have a precise timing for a process which is activated only for a short interval [39]. Another example is the rhythmic activity of the nervous system [40], which rests on the ability of individual neural cells to become excited only if the external stimulus exceeds a threshold value. In contrast, the biological importance of oscillations is less apparent in the case of metabolic processes such as glycolysis. One could argue, however, that oscillations in ATP arising from the phosphofructokinase control provide a periodic input for the energy-requiring mechanisms of synthesis of the large molecules, and therefore that they trigger indirectly the timing of <u>all</u> cellular processes, including, for example, mitosis. Further study is necessary for a more complete understanding of this important point.

CHAPTER VI: THE CONTROL OF PROTEIN BIOSYNTHESIS

A. Introduction. [41]

The second general type of control mechanism in living cells is the regulation of the synthesis of proteins. A major function of proteins is to act as enzymes which catalyze various reactions. Two mechanisms exist which control the amount of enzymes present in cells, repression and induction. Since proteins are fairly stable molecules (lifetimes on the order of several minutes) and catalysis is a very fast process, there may occur a situation in which the amount of enzyme is too great. In response to such an excess in the cell, a metabolite (usually related to the substrate or product of the enzymatic reaction) causes the repression of the synthesis of the particular enzyme. In other circumstances, the organism may be subjected to a change of environment. New types of enzymes are needed then to catalyze reactions which utilize as substrates molecules found in the new environment. In such a situation, induction of the synthesis of a new enzyme may be initiated by the new substrate itself.

Before we go into the details of specific examples of such mechanisms it will be helpful to describe briefly the genetic aspects of the synthesis of proteins.

B. The Biosynthesis of Proteins.

Nucleic acids and proteins are two types of biological macromolecules (heavy polymers consisting of a large number of units). Both types of molecules are linear and they are related in a linear manner as will be seen shortly.

Nucleic acids consist of a backbone of phosphate and sugar--deoxyribose in the case of deoxyribonucleic acid (DNA) and ribose in the case of ribonucleic acid (RNA). To each sugar is attached one of four bases: adenine (A), guanine (G), cytosine (C), or thymine (T) in DNA and A, G, C or uracil (U) in RNA. The classical work of Watson and Crick [42] elucidated the complementarity of bases: A forms hydrogen bonds with U or T, G with C. DNA exists as a double helix, two strands held together by the binding of complementary base pairs; RNA usually exists in a single stranded form.

Proteins are polymers of twenty different amino acids linked by peptide bonds: $-\overset{\displaystyle O}{\overset{\|}{C}}-NH-$. The specific enzymatic properties of a protein are a direct consequence of the spatial configuration of the molecule under physiological conditions. In order that each molecule of a protein species has the same amino acid sequence and thus the same biological activity, an invariant method exists for building each protein macromolecule from its constituent amino acids. This process is part of the central dogma of molecular biology and is often summarized: DNA→RNA→protein. By virtue of its double helical structure, DNA is a very stable molecule. Information specifying the amino acid sequence of a protein is found coded in part of a DNA molecule (gene or operon). A molecule of messenger RNA (mRNA) complementary to the DNA gene is transcribed from the gene by means of a specific enzyme. The resulting mRNA molecule, in association with large ribonucleoprotein complexes called ribosomes, is used as a template in the synthesis of the protein. Each triplet of bases (a codon, of which 4^3 or 64 are possible) on the mRNA signifies a particular amino acid. Amino acid molecules are supplied by means of transfer RNA (tRNA) molecules. Each tRNA species contains an anticodon (complementary in base sequence to the codon on the mRNA) and a site for binding the correct amino acid. As tRNA's and mRNA's match by means of codon-anticodon interactions, amino acids are brought into proximity. Peptide bonds are then formed enzymatically between adjacent amino acids. Fig. 15 is a useful schematic representation of the process of protein synthesis; the correspondence between codons and amino acids appears in Table 2.

C. The Jacob-Monod Model for Biosynthetic Control.

In contrast to the control of metabolite synthesis (Ch. V), the regulation of protein biosynthesis typically involves the interaction of pathways corresponding to more than one enzyme. Beginning with this observation, Jacob and Monod [43] have proposed a number of ingenious models for biosynthetic control. The key features of the Jacob-Monod models are (1) repression of enzyme synthesis by influence on the genetic material of products of the metabolic action of the enzyme itself and/or (2) induction of enzyme synthesis due to activation by metabolites of part of the genetic material. In this section we discuss a simple such model involving interaction between two enzymatic pathways [44].

Fig. 15 Synthesis of a Protein Molecule.* Each codon of the messenger RNA is "read" in sequence by the ribosome moving along the mRNA chain. An appropriate molecule of transfer RNA then binds temporarily with the codon, and its amino acid is bound enzymatically to the chain which is being synthesized. The ribosome has two binding sites for molecules of tRNA: one (A) for positioning a newly arrived tRNA molecule and another (B) for holding the growing polypeptide chain.**

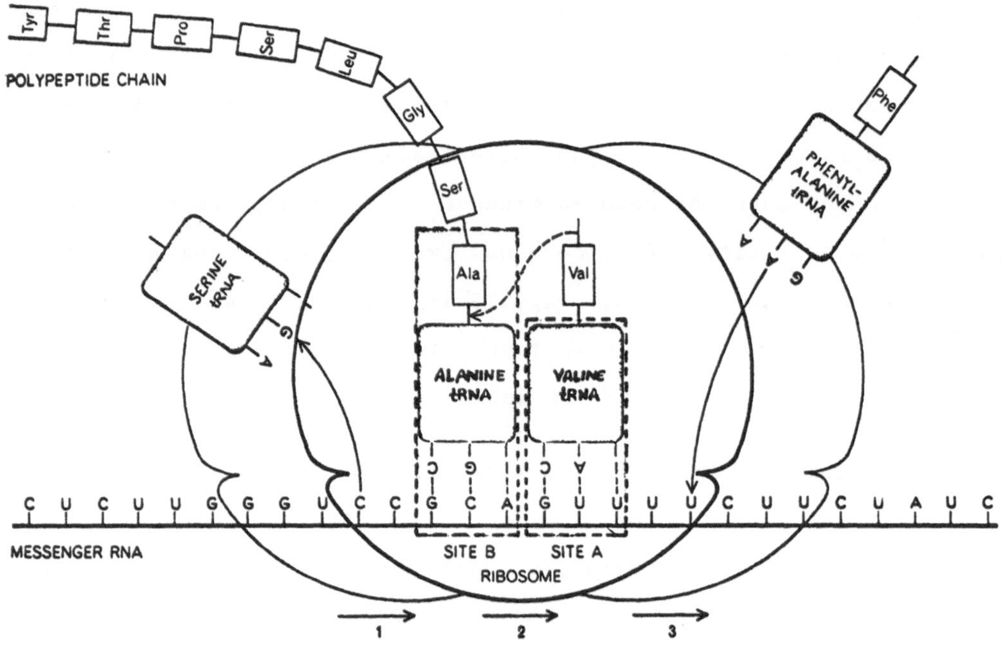

---

*From "The Genetic Code: III" by F. H. C. Crick. Copyright © 1966 by Scientific American, Inc. All rights reserved.

**Codon-anticodon pairing at the third base position is more complicated than usual rules of complementarity indicate. Here I stands for inosine (closely resembling G) which, according to the "wobble hypothesis" of Crick, may form bonds with U, C, and A (see Ch. 22 of Ref. [41]).

Table 2. The Genetic Code

The sixty-four possible triplets for coding amino acids are displayed in compact form. Each numerical entry in the table denotes the amino acid corresponding to a particular triplet (codon):

1. Alanine	8. Glycine	15. Proline
2. Arginine	9. Histidine	16. Serine
3. Asparagine	10. Isoleucine	17. Threonine
4. Aspartic acid	11. Leucine	18. Tryptophan
5. Cysteine	12. Lysine	19. Tyrosine
6. Glutamic acid	13. Methionine	20. Valine
7. Glutamine	14. Phenylalanine	21. End chain

For example the triplet CAG stands for No. 7. This implies that cytosine-adenine-guanine codes glutamine.

1st letter	2nd letter: U	C	A	G	3rd letter
U	14	16	19	5	U
	14	16	19	5	C
	11	16	21	21	A
	11	16	21	18	G
C	11	15	9	2	U
	11	15	9	2	C
	11	15	7	2	A
	11	15	7	2	G
A	10	17	3	16	U
	10	17	3	16	C
	10	17	12	2	A
	13	17	12	2	G
G	20	1	4	8	U
	20	1	4	8	C
	20	1	6	8	A
	20	1	6	8	G

The model is represented schematically in Fig. 16. The regulatory gene RG of each enzyme system provides a repressor molecule R which when activated by the product P of the other pathway (to give Re) can combine with the operator gene O in such a way that transcription by gene G of the corresponding mRNA is blocked. The system is open to reservoirs of substrates S and, in addition to regulatory action the product P_1, (but not P_2, for simplicity) participates in further metabolic reactions having product F_1.*

If the inactive repressor R is assumed to combine with two product molecules to form Re, then the overall scheme expressed as a set of chemical reactions is

$$R_1 + 2P_2 \underset{k_{-1}}{\overset{k_1}{\rightleftharpoons}} Re_1 \qquad R_2 + 2P_1 \underset{k_{-8}}{\overset{k_8}{\rightleftharpoons}} Re_2$$

$$G_1^+ + Re_1 \underset{k_{-2}}{\overset{k_2}{\rightleftharpoons}} G_1^- \qquad G_2^+ + Re_2 \underset{k_{-9}}{\overset{k_9}{\rightleftharpoons}} G_2^-$$

$$G_1^+ + N_1 \xrightarrow{k_3} G_1^+ + X_1 \qquad G_2^+ + N_2 \xrightarrow{k_{10}} G_2^+ + X_2$$

$$X_1 \xrightarrow{k_4} C_1 \qquad X_2 \xrightarrow{k_{11}} C_2 \tag{6.1a}$$

$$M_1 + X_1 \xrightarrow{k_5} E_1 + X_1 \qquad M_2 + X_2 \xrightarrow{k_{12}} E_2 + X_2$$

$$E_1 \xrightarrow{k_6} D_1 \qquad E_2 \xrightarrow{k_{13}} D_2$$

$$E_1 + S_1 \xrightarrow{k_7} P_1 + E_1 \qquad E_2 + S_2 \xrightarrow{k_{14}} P_2 + E_2$$

$$P_1 \underset{k_{-15}}{\overset{k_{15}}{\rightleftharpoons}} F_1 \ . \tag{6.1b}$$

---

*Such a double role for product is not at all unusual. In Escherichia coli grown in lactose the inducer of the enzyme β-galactosidase is allolactose [45], obtained from lactose by a transgalactosidation reaction for which β-galactosidase itself is responsible. In addition allolactose, like lactose, is hydrolyzed by the enzyme.

Here X denotes mRNA, C and D the decay products for mRNA and enzyme, N and M the building blocks for mRNA and enzyme, and the probabilities for open (+) and closed (-) gene satisfy the normalization condition $G_i^+ + G_i^- = 1$.

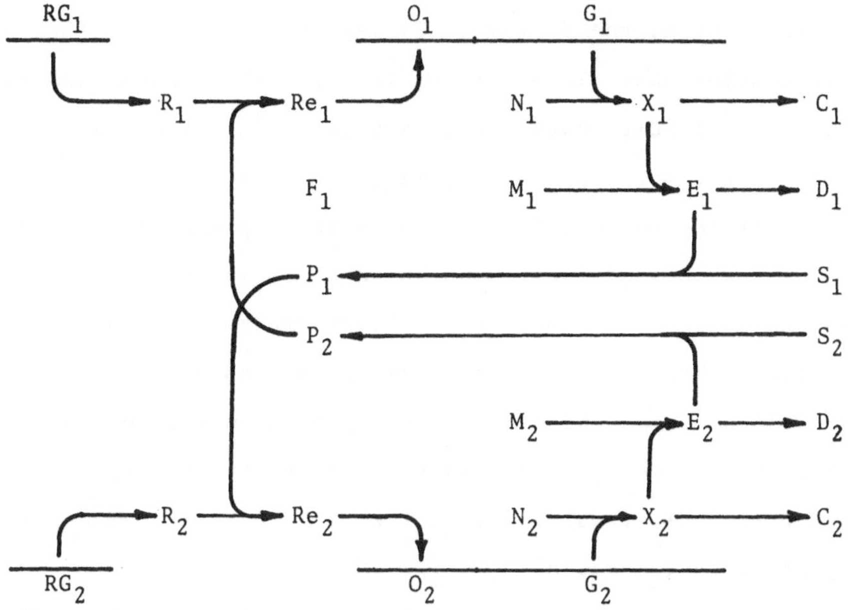

Figure 16

If we now make the natural assumption that syntheses of mRNA and enzyme constitute the rate-determining processes in scheme (6.1), then some straightforward manipulations yield a set of four nonlinear rate equations for mRNA's and enzymes:

$$\begin{aligned}
\frac{dE_1}{dt} &= k_5 M_1 X_1 - k_6 E_1 \\
\frac{dE_2}{dt} &= k_{12} M_2 X_2 - k_{13} E_2 \\
\frac{dX_1}{dt} &= -k_4 X_1 + f[E_2^2] \\
\frac{dX_2}{dt} &= -k_{11} X_2 + f[(E_1 + \alpha F_1)^2] \ ,
\end{aligned} \tag{6.2}$$

where the function f and parameter α depend on various rate constants and reservoir concentrations.

To study the stationary state behavior of this system we set the time-derivatives to zero and after some elementary algebra obtain for the steady state equation a quintic in E_1. For appropriate values of the parameters, then, the system (6.2) may possess at most five possible steady state solutions. In practice, the system is found (numerically) to exhibit three simultaneous steady states for a narrow range of concentrations of the decay product F_1 [44]. A typical result appears in Fig. 17. According to an infinitesimal stability analysis of the system (6.2), the upper and lower branches are stable while the middle branch is unstable. This implies an abrupt transition from the low-E branch to the higher one as F_1 is increased beyond 0.5 (arrow, Fig. 17). For other values of the parameters the region of multiple steady states no longer exists. The E_1-F_1 relationship is given then by a sigmoidal curve, and a gradual transition from low to high E_1 states occurs continuously as F_1 is increased.

For appropriate choices of various parameters, therefore, two distinct types of transitions are possible. Both are seen to involve a switch between widely different enzyme concentrations. Accompanying either transition, therefore, is a radical change in functional behavior without any alteration of the basic genetic information available to the system. In addition, both types of transition exhibit great sensitivity to external regulation (e.g., via change in F_1). Despite these similarities, however, one expects that transitions of the discontinuous type (Fig. 17) may well be favored biologically due to greater efficiency with respect to changes in the environment. Indeed, the existence of such transitions in growing populations of bacteria has been known for some time [46]. Specifically, the induction of β-galactosidase in Escherichia coli is observed to be an "all-or-none" phenomenon [46,47]. Theoretical results agreeing with this conclusion have been obtained recently for a simple model incorporating the principal features of the lactose operon system in E. coli [48]. It is important to note that such discontinuous transitions between multiple steady states are inherently far-from-equilibrium phenomena due both to the particular (nonlinear) irreversible processes occurring in the system and to an asymmetry in the system maintained by specific nonequilibrium coupling to the environment (here, the reservoirs of S_1, S_2, and F_1).

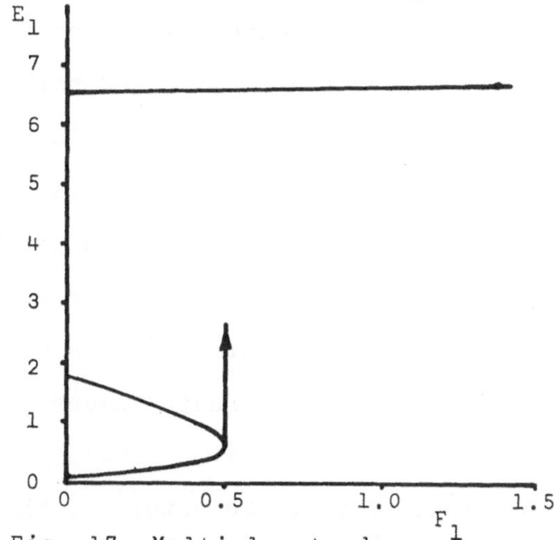

Fig. 17. Multiple steady states for model (6.2).

In the model discussed here the essential nonlinearity is introduced by the assumption that two product molecules P combine with R to form the active repressor Re. Similar results involving transitions between multiple steady states obtain when this nonlinearity is replaced or supplemented by others due to additional interactions between the pathways. Hence for a wide variety of model enzyme systems one can find multiple steady states, each corresponding to the functioning of one pathway and inhibition of others. In particular one may imagine an extreme situation like that shown in Fig. 17. Here a reverse transition to the low-E branch does not exist (i.e., the hysteresis loop closes in the $F_1 < 0$ domain, so the high-E branch is stable for all allowable values of F_1). The forward transition appears then to be irreversible. Hence, if at some time one pathway gets a head start due to temporary metabolic advantage (e.g., a spontaneous excess of F_1), all other pathways may be inhibited permanently even after the temporary constraint is released. On the cellular level such irreversible changes of functional behavior in response to "signals" from the environment may be important in discussions of cell differentiation, for example.

In view of the far-reaching consequences of the latter possibility especially, it is appropriate to conclude this review with two remarks concerning the application of simple mathematical models to actual biological systems: First, it is obvious that simple models of such systems necessarily treat only a few of the many degrees of freedom available to the particular subsystem under consideration. In the model studied here, for example, the switch between pathways is accomplished by varying the concentration of the metabolic product F_1 while all other reservoir concentrations and parameters are fixed. It is not difficult to verify, however, that similar effects may be obtained by varying instead certain kinetic constants (e.g., via changes in pH) or other substrate concentrations. In particular, by direct or indirect coupling (not included in the model) to other processes in a cell, for example, some of the constant variables may be altered by the forward transition in such a way that reattainment of the initial low-E_1 branch is no longer prevented [49]. Finally, a more difficult question concerns the possibility that the nature of the steady states predicted by the macroscopic rate equations may be incorrect in the coexistence regions of multiple steady states, particularly when the relatively small populations of various enzymes, RNA's, etc. appropriate to single cells are considered. For discussion of this problem see Secs. III-C and IV-E and Refs. [13,15].

APPENDIX A

The following are general reviews or articles discussing a number of examples related to chemical instabilities:

A. General.

1. G. Nicolis: "Stability and Dissipative Structures in Open Systems Far From Equilibrium," Adv. Chem. Phys. 19, 209-324 (1971). [Same as Ref. [8]; Especially good for mathematical discussion of stability theory and of chemical oscillations in conservative and non-conservative systems.]

2. I. Prigogine and R. Lefever: "Thermodynamics, Structure, and Dissipation," Experientia Suppl. 18, 101-126 (1971). [Experimental and theoretical discussion of membrane-related instabilities.]

3. G. Nicolis and J. Portnow: "Chemical Oscillations," Chem. Revs. (in press, 1973). [Same as Ref. [10]; Comprehensive review of theory and experiment.]

B. Biological Orientation.

4. I. Prigogine, G. Nicolis, and A. Babloyantz: "Thermodynamics of Evolution," Phys. Today 25 (No. 11) 23 and (No. 12) 38 (1972). [Same as Ref. [1a]; Prebiotic evolution of macromolecules.]

5. M. Eigen: "Selforganization of Matter and the Evolution of Biological Macromolecules," Naturwissenschaften 58, 465-523 (1971). [Stochastic approach to prebiotic evolution.]

6. I. Prigogine and G. Nicolis: "Biological Order, Structure and Instabilities," Quart. Revs. Biophys. 4, 107-148 (1971). [Same as Ref. [16]; Discusses dissipative structures in the context of a number of biological phenomena.]

APPENDIX B

A Single Irreversible Process: Thermal Conduction

Consider an isotropic solid subject to a time-independent non-uniform temperature distribution at its boundaries (neglect thermal expansion). According to Eq. (1.28), the local entropy production is then

$$\sigma = \underset{\sim}{W} \cdot \text{grad} \frac{1}{T} , \qquad \text{(B-1)}$$

where the heat flow is assumed to obey a linear law

$$\underset{\sim}{W} = L \text{ grad} \frac{1}{T} . \qquad \text{(B-2)}$$

The phenomenological coefficient is related to the thermal conductivity by $L = \lambda T^2$, and is assumed to be approximately constant throughout the system, depending only on the overall equilibrium temperature. The temperature distribution corresponding to minimum entropy production follows from the variational equation

$$\delta P = \delta \int \sigma dV = \delta \{L \int (\text{grad} \frac{1}{T})^2 dV\} = 0 . \qquad \text{(B-3)}$$

The Euler-Lagrange equation for this problem, for temperature variations δT which vanish at the boundaries, is

$$\text{div grad} \frac{1}{T} = 0 , \qquad \text{(B-4a)}$$

or, from Eq. (B-2),

$$\text{div } \underset{\sim}{W} = 0 . \qquad \text{(B-4b)}$$

According to Eq. (1.18), the energy balance equation for this system takes the form

$$-\text{div } W = \rho \frac{\partial e}{\partial t} = \rho c_V \frac{\partial T}{\partial t} , \qquad \text{(B-5)}$$

where the last equality introduces the specific heat at constant volume, c_V. Inserting the result (B-4b) into this equation yields

$$\frac{\partial e}{\partial t} = 0 \quad \text{or} \quad \frac{\partial T}{\partial t} = 0 , \qquad \text{(B-6)}$$

which verifies that the steady state is indeed characterized by minimum entropy production.

Coupled Irreversible Processes

Consider now two fluid subsystems, seperately at equilibrium at different temperatures, which are separated by a membrane through which energy and matter may flow. There are now two generalized forces: X_{th} corresponding to the fixed temperature difference, and X_m corresponding to the difference of chemical potentials between the two subsystems. The latter force is unrestrained, so its conjugate matter flow J_m will vanish at the steady state. The heat flow J_{th}, as well as the entropy production, will then approach constant values, and the state variables will become time-independent. The local entropy production for this system has the form (1.30)

$$\sigma = J_{th} X_{th} + J_m X_m > 0 \,. \qquad \text{(B-7)}$$

If linear phenomenological laws hold

$$J_{th} = L_{11} X_{th} + L_{12} X_m$$
$$J_m = L_{21} X_{th} + L_{22} X_m \,, \qquad \text{(B-8)}$$

then the entropy source becomes

$$\sigma = L_{11} X_{th}^2 + 2L_{21} X_{th} X_m + L_{22} X_m^2 \,, \qquad \text{(B-9)}$$

where Onsager's reciprocal relation $L_{12} = L_{21}$ has been used. With the system constrained by a fixed X_{th}, σ is varied with respect to the variable force X_m:

$$\frac{\partial\sigma}{\partial X_m} = 2(L_{21} X_{th} + L_{22} X_m) = 2 J_m = 0 \,. \qquad \text{(B-10)}$$

Hence again the equivalence of minimum entropy production and the steady state condition (i.e., $J_m = 0$) is established.

Selecting New Forces and Flows: Chemical Reactions

In both preceding examples the natural choice of generalized forces was compatible with the imposed constraints. In many instances, however, this may not be the case. Consider a simple linear scheme of chemical reactions:

$$A \underset{k_{-1}}{\overset{k_1}{\rightleftharpoons}} X \underset{k_{-2}}{\overset{k_2}{\rightleftharpoons}} B \,. \qquad \text{(B-11)}$$

The system is maintained homogeneous and isothermal with time-independent concentrations of A and B. Again we write the local entropy production from Eq. (1.28):

$$T\sigma = w_1 A_1 + w_2 A_2 , \qquad \text{(B-12)}$$

where

$$w_1 = k_1 A - k_{-1} X , \quad w_2 = k_2 X - k_{-2} B , \qquad \text{(B-13)}$$

and for an ideal reacting mixture the affinities (in units of RT, where R is the gas constant) take the form

$$A_1 = \ln \frac{k_1 A}{k_{-1} X} , \quad A_2 = \ln \frac{k_2 X}{k_{-2} B} . \qquad \text{(B-14)}$$

Writing rates in terms of affinities yields the following linear laws:

$$w_1 = k_1 A(1 - \frac{k_{-1} X}{k_1 A}) = k_1 A(1-e^{-A_1}) \approx k_1 A A_1$$

$$\text{(B-15)}$$

$$w_2 = k_{-2} B(\frac{k_2 X}{k_{-2} B} - 1) = k_{-2} B(e^{A_2}-1) \approx k_{-2} B A_2 ,$$

provided the system is near equilibrium (i.e., provided $A_1, A_2 << 1$). Here the cross coefficients are zero, and for fixed A, B the direct coefficients are constant.

Although it is clear from Eq. (B-14) that neither affinity can be related directly to the constraints, we do find that the quantity

$$A_1 + A_2 = \ln \frac{k_1 k_2 A}{k_{-1} k_{-2} B} \qquad \text{(B-16)}$$

is indeed constant for fixed A, B. Hence it is convenient to define new forces Z_1, Z_2

$$\gamma Z_1 = A_1 + A_2 , \quad \gamma Z_2 = A_1 - A_2 \qquad \text{(B-17a)}$$

and new conjugate flows v_1, v_2

$$\gamma v_1 = w_1 + w_2 , \quad \gamma v_2 = w_1 - w_2 , \qquad \text{(B-17b)}$$

where $\gamma = 2^{1/2}$. It is easy to verify that the local entropy production is unchanged by the transformation and retains the characteristic form (B-12):

$$T\sigma = v_1 Z_1 + v_2 Z_2 . \qquad \text{(B-18)}$$

The key feature of this expression is the decomposition into an unrestrained force Z_2 and a force Z_1 which is fixed by the external constraints.

If one writes linear laws for the new flow variables,

$$V_1 = L_{11} Z_1 + L_{12} Z_2 \quad , \quad V_2 = L_{21} Z_1 + L_{22} Z_2 \,, \qquad \text{(B-19)}$$

then one finds from Eqs. (B-15) and (B-17) that

$$L_{11} = L_{22} = k_1 A + k_{-2} B \quad , \quad L_{12} = L_{21} = k_1 A - k_{-2} B \,. \qquad \text{(B-20)}$$

As we require, therefore, the new phenomenological coefficients are constant satisfying Onsager's reciprocal relations.

Minimizing the entropy production with respect to the unrestrained force Z_2 yields, from Eqs. (B-18) to (B-20),

$$T\frac{\partial \sigma}{\partial Z_2} = 2(L_{21} Z_2 + L_{22} Z_2) = 2 v_2 = 0 \,. \qquad \text{(B-21)}$$

To see that the condition $v_2 = 0$ indeed corresponds to a steady state, one has only to note that for the scheme (B-11) the time rate of change of the concentration of intermediate X is simply

$$\dot{X} = w_1 - w_2 = \gamma v_2 \,.$$

REFERENCES

[1] For an overview and recent developments in a biological context, see I. PRIGOGINE, G. NICOLIS, and A. BABLOYANTZ: Physics Today 25 (No. 11) 23 and (No. 12) 38 (1972); I. PRIGOGINE and G. NICOLIS: Quart. Rev. Biophys. 4, 107 (1971).

[2] P. GLANSDORFF AND I. PRIGOGINE: Thermodynamic Theory of Structure, Stability and Fluctuations (Wiley-Interscience, New York, 1971).

[3] (a) For a comprehensive discussion of linear irreversible thermodynamics within the framework of the local equilibrium assumption, see S. R. DEGROOT and P. MAZUR: Non-Equilibrium Thermodynamics (North Holland, Amsterdam, 1962); (b) I. PRIGOGINE: Introduction to Thermodynamics of Irreversible Processes, 3rd Ed. (Wiley-Inter-Science, New York, 1967); (c) A. KATCHALSKY and P. F. CURRAN: Nonequilibrium Thermodynamics in Biophysics (Harvard Univ. Press, Cambridge, Mass., 1965).

[4] I. PRIGOGINE: Physica 14, 272 (1949); G. NICOLIS, J. WALLENBORN, and M. G. VELARDE: Physica 43, 263 (1969).

[5] L. ONSAGER: Phys. Rev. 37, 405 and 38, 2265 (1931).

[6] Indeed, in his original work [5] Onsager presented a principle of least dissipation of energy for steady nonequilibrium states. Anticipating later extension to nonlinear irreversible processes, we consider instead an approach based on the theorem of minimum entropy production (due to Prigogine: see Refs. [2] or [3] for details).

[7] P. GLANSDORFF AND I. PRIGOGINE: Physica 20, 773 (1954); ibid, 30, 351 (1964).

[8] G. NICOLIS: Adv. Chem. Phys. 19, 209 (1971).

[9] For details and proofs see, e.g., N. MINORSKY: Nonlinear Oscillations (Van Nostrand, Princeton, N. J., 1962).

[10] G. NICOLIS AND J. PORTNOW: "Chemical Oscillations", Chem. Revs. (to appear, 1973).

[11] G. NICOLIS: to be published (1973).

[12] M. HERSCHKOWITZ-KAUFMAN AND J. K. PLATTEN: Bull. Cl. Sci. Acad. Roy. Belg. 52, 26 (1971); M. HERSCHKOWITZ-KAUFMAN and G. NICOLIS: J. Chem. Phys. 56, 1890 (1972).

[13] J. S. TURNER: Physics Letters 44A, 395 (1973); Bull. Math. Biophys. (Dec., 1973).

[14] Y. KOBATAKE: Physica 48, 301 (1970).

[15] J. S. TURNER: in Membranes, Dissipative Structures and Evolution, proceedings of a workshop, Brussels, 1972 (to appear, 1974).

[16] D. A. MCQUARRIE: Stochastic Approach to Chemical Kinetics, Vol. 8 of Suppl. Rev. Ser. in Applied Probability (Methuen, London, 1967).

[17] G. NICOLIS AND A. BABLOYANTZ: J. Chem. Phys. 51, 2632 (1969).

[18] G. NICOLIS AND I. PRIGOGINE: Proc. Nat. Acad. Sci. (USA) 68, 2102 (1971); G. NICOLIS: J. Stat. Phys. 6, 195 (1972).

[19] A general introductory reference for this chapter is A.L. LEHNINGER: Bioenergetics: The Molecular Basis of Biological Energy Transformations, 2nd Ed. (Benjamin, Menlo Park, Cal., 1971).

[20] For an excellent discussion and a complete bibliography up to 1970, see B. HESS and A. BOITEUX: Ann. Rev. Biochem. 40, 237 (1971).

[21] L. N. M. DUYSENS AND J. AMESZ: Biochim. Biophys. Acta 24, 19 (1957).

[22] A. GHOSH AND B. CHANCE: Biochem. Biophys. Res. Comm. 16, 174 (1964); B. CHANCE, R. W. EASTABROOK, and A. GHOSH: Proc. Nat. Acad. Sci. (USA) 51, 1244 (1964).

[23] B. CHANCE, B. HESS, AND A. BETZ: Biochem. Biophys. Res. Comm. 16, 182 (1964); K. PYE and B. CHANCE: Proc. Nat. Acad. Sci. (USA) 55, 888 (1966).

[24] K. H. IBSEN AND K. W. SCHILLER: Biochim. Biophys. Acta 131, 405 (1967).

[25] R. FRENKEL: Biochem. Biophys. Res. Comm. 21, 497 (1965).

[26] A. BOITEUX AND B. HESS: F.E.B.S. Meet. 4th, Oslo Abstr. #398 (1967).

[27] B. HESS AND A. BOITEUX: in Regulatory Functions of Biological Membranes, J. Jarnefelt, Ed. (Elsevier, New York, 1967).

[28] B. HESS, A. BOITEUX, AND J. KRUGER: Advan. Enzyme Regul. 7, 149 (1969); B. CHANCE, B. HESS, J. HIGGINS, and K. PYE: Biochemical Oscillations (Academic Press, New York, in press).

[29] E. E. SEL'KOV: Europ. J. Biochem. 4, 79 (1968).

[30] E. E. SEL'KOV: Molec. Biol. (USSR) 2, 252 (1968); V.A. SAMOILENKO, E. E. SEL'KOV, and V. C. SAVCHUK: in Oscillatory Processes in Biological and Chemical Systems Vol. II (Nauka, Puschino on Oka, 1967), p. 54.

[31] J. HIGGINS: Proc. Nat. Acad. Sci. (USA) 51, 989 (1964).

[32] For details see Ref. [29] and Ref. 6-9, 24-30 therein.

[33] A. GOLDBETER AND R. LEFEVER: Biophys. J. 12, 1302 (1972); Proc. 8th F.E.B.S. Meeting (Amsterdam, 1972).

[34] Experimental studies indicate that phosphofructokinase is a tetramer. Qualitatively, however, one expects that the behavior of the PFK scheme is adequately approximated by the dimer model considered in this section.

[35] J. MONOD, J. WYMAN, AND J.-P. CHANGEUX: J. Mol. Biol. 12, 88 (1965).

[36] A. GOLDBETER: Proc. Nat. Acad. Sci. (USA) (in press, 1973).

[37] B. HESS: Private communication to G. Nicolis.

[38] C. F. WALTER: J. Theor. Biol. 27, 259 (1970).

[39] B. C. GOODWIN: Temporal Organization in Cells (Academic Press, New York, 1963).

[40] J. COWAN: in Neural Networks, E. R. Caianiello, Ed. (Springer-Verlag, Berlin, 1968).

[41] Useful introductory material for this chapter appears in a volume of readings from Scientific American: The Molecular Basis of Life, R. H. Haynes and P. C. Hanawalt, Eds. (W. H. Freeman, San Francisco, 1968). See especially chapters 20 [F. H. C. Crick: Sci. Am. (Oct., 1962)], 21 [M. W. Nirenberg: Sci. Am. (Mar., 1963)], 22 [F. H. C. Crick: Sci. Am. (Oct., 1966)], and 25 [J.-P. Changeux: Sci. Am. (Apr., 1965)].

[42] J. D. WATSON AND F. H. C. CRICK: Nature 171, 737 and 964 (1953).

[43] F. JACOB AND J. MONOD: J. Mol. Biol. 3, 318 (1961).

[44] A. BABLOYANTZ AND G. NICOLIS: J. Theor. Biol. 34, 185 (1972).

[45] C. BURSTEIN, M. COHN, A. KEPES, AND J. MONOD: Biochim. Biophys. Acta 95, 634 (1965).

[46] A. NOVICK AND M. WEINER: in Symposium on Molecular Biology (Univ. of Chicago Press, 1959), p. 78.

[47] F. JACOB AND J. MONOD: Cold Spring Harb. Symp. Quant. Biol. 26, 193 (1961).

[48] A. BABLOYANTZ AND M. SANGLIER: F.E.B.S. Letters 23, 364 (1972).

[49] A simple example in which the initial regime may be reestablished continuously after a discontinuous transition is discussed by the author in Refs. [13] and [15] (See also Sec. III-C).

Lecture Notes in Physics

Bisher erschienen / Already published

Vol. 1: J. C. Erdmann, Wärmeleitung in Kristallen, theoretische Grundlagen und fortgeschrittene experimentelle Methoden. 1969. DM 20,–

Vol. 2: K. Hepp, Théorie de la renormalisation. 1969. DM 18,–

Vol. 3: A. Martin, Scattering Theory: Unitarity, Analyticity and Crossing. 1969. DM 16,–

Vol. 4: G. Ludwig, Deutung des Begriffs physikalische Theorie und axiomatische Grundlegung der Hilbertraumstruktur der Quantenmechanik durch Hauptsätze des Messens. 1970. DM 28,–

Vol. 5: M. Schaaf, The Reduction of the Product of Two Irreducible Unitary Representations of the Proper Orthochronous Quantummechanical Poincaré Group. 1970. DM 16,–

Vol. 6: Group Representations in Mathematics and Physics. Edited by V. Bargmann. 1970. DM 24,–

Vol. 7: R. Balescu, J. L. Lebowitz, I. Prigogine, P. Résibois, Z. W. Salsburg, Lectures in Statistical Physics. 1971. DM 18,–

Vol. 8: Proceedings of the Second International Conference on Numerical Methods in Fluid Dynamics. Edited by M. Holt. 1971. DM 28,–

Vol. 9: D. W. Robinson, The Thermodynamic Pressure in Quantum Statistical Mechanics. 1971. DM 16,–

Vol. 10: J. M. Stewart, Non-Equilibrium Relativistic Kinetic Theory. 1971. DM 16,–

Vol. 11: O. Steinmann, Perturbation Expansions in Axiomatic Field Theory. 1971. DM 16,–

Vol. 12: Statistical Models and Turbulence. Edited by M. Rosenblatt and C. Van Atta. 1972. DM 28,–

Vol. 13: M. Ryan, Hamiltonian Cosmology. 1972. DM 18,–

Vol. 14: Methods of Local and Global Differential Geometry in General Relativity. Edited by D. Farnsworth, J. Fink, J. Porter and A. Thompson. 1972. DM 18,–

Vol. 15: M. Fierz, Vorlesungen zur Entwicklungsgeschichte der Mechanik. 1972. DM 16,–

Vol. 16: H.-O. Georgii, Phasenübergang 1. Art bei Gittergasmodellen. 1972. DM 18,–

Vol. 17: Strong Interaction Physics. Edited by W. Rühl and A. Vancura. 1973. DM 28,–

Vol. 18: Proceedings of the Third International Conference on Numerical Methods in Fluid Mechanics, Vol. I. Edited by H. Cabannes and R. Temam. 1973. DM 18,–

Vol. 19: Proceedings of the Third International Conference on Numerical Methods in Fluid Mechanics, Vol. II. Edited by H. Cabannes and R. Temam. 1973. DM 26,–

Vol. 20: Statistical Mechanics and Mathematical Problems. Edited by A. Lenard. 1973. DM 22,–

Vol. 21: Optimization and Stability Problems in Continuum Mechanics. Edited by P. K. C. Wang. 1973. DM 16,–

Vol. 22: Proceedings of the Europhysics Study Conference on Intermediate Processes in Nuclear Reactions. Edited by N. Cindro, P. Kulišić and Th. Mayer-Kuckuk. 1973. DM 26,–

Vol. 23: Nuclear Structure Physics. Proceedings of the Minerva Symposium on Physics. Edited by U. Smilansky, I. Talmi, and H. A. Weidenmüller. 1973. DM 26,–

Vol. 24: R. F. Snipes, Statistical Mechanical Theory of the Electrolytic Transport of Non-electrolytes. 1973. DM 20,–

Vol. 25: Constructive Quantum Field Theory. The 1973 "Ettore Majorana" International School of Mathematical Physics. Edited by G. Velo and A. Wightman. 1973. DM 26,–

Vol. 26: A. Hubert, Theorie der Domänenwände in geordneten Medien. 1974. DM 28,–

Vol. 27: R. Kh. Zeytounian, Notes sur les Ecoulements Rotationnels de Fluides Parfaits. 1974. DM 28,–

Vol. 28: Lectures in Statistical Physics. Edited by W. C. Schieve and J. S. Turner. 1974. DM 24,–

Selected Issues from

Lecture Notes in Mathematics

Selected Issues from

Lecture Notes in Physics